Observation Exercises in Astronomy

Lauren Jones
Columbus State Community College

Addison-Wesley
Boston Columbus Indianapolis New York San Francisco Upper Saddle River
Amsterdam Cape Town Dubai London Madrid Milan Munich Paris Montréal Toronto
Delhi Mexico City São Paulo Sydney Hong Kong Seoul Singapore Taipei Tokyo

Publisher: Jim Smith
Executive Editor: Nancy Whilton
Director of Development: Michael Gillespie
Editorial Manager: Laura Kenney
Project Editor: Steven Le
Senior Marketing Manager: Kerry Chapman
Managing Editor: Corinne Benson
Production Supervisor: Mary O'Connell
Compositor: Progressive Information Technologies
Production Service: Progressive Publishing Alternatives
Cover Design: Marilyn Perry
Manufacturing Buyer: Jeffrey Sargent
Manager, Cover Visual Research & Permissions: Karen Sanatar
Cover Printer: Phoenix Color
Printer and Binder: Edwards Brothers, Inc.

Cover Photo Credit: NGC 602 and Beyond, NASA, ESA, and the Hubble Heritage Team (STScI/AURA)—ESA/Hubble Collaboration

Credits and acknowledgments borrowed from other sources and reproduced, with permission, in this textbook appear on the appropriate page within the text

ISBN 10: 0-321-63812-3; ISBN 13: 978-0-321-63812-0

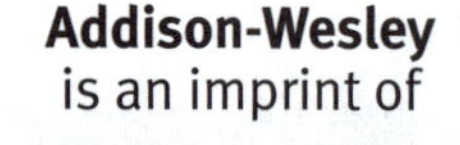

www.pearsonhighered.com

1 2 3 4 5 6 7 8 9 10—EB—13 12 11 10

Table of Contents

Acknowledgments

The author would like to thank her family for being so supportive during the writing of this book. Also, I would like to thank all the people who, in one way or another, made this book possible (in no particular order): Richard Elston, Henry Kandrup, Haywood Smith, Steve Gottesman, Debbie Elmegreen, Cindy Schwarz, William Sweeney, Joseph Profeta, Gene Byrd, William Keel, Anatoly Zasov, Andrei Rosterguyev, Nikolai Shakura, Alex Smith, Francisco Reyes, Timothy Slater, Ed Prather, Larry Marschall, Dick Cooper, Jim Brown, John Caraher, Esther Hopkins, Adrienne Carr, all my students over the years, and my mother, Constance Jones.

Introduction

Welcome to the first attempt at a suite of exercises for students of Astronomy 101 that both incorporates the new technologies of the 21st century and addresses the needs of the Astronomy 101 student population. These exercises are written for the typical student population for this course. These students are mostly nonscience majors who have not taken many college math courses and may have significant math fear. The goal of this lab manual is to work with this population of students—who are not likely to choose astronomy or physics for a major—to help make them more comfortable with science and math, to experience how science is done, to understand how scientists know what they know, and to understand why the pursuit of science is important to society.

The order of these exercises is intentional. They follow a traditional one-semester introductory astronomy syllabus. Most importantly, they build on one another and introduce the process of science as well as the use of different software programs (including planetarium software). Exercises toward the end of this manual involve the application of mathematical formulae using Excel spreadsheets (which does not require the student to actually do math, but does require them to know how to use Excel), as well as graphing with Excel.

If it suits your course to do these exercises in a different sequence, it is highly recommended that time be spent reviewing the content of the previous exercises to make sure your students will be prepared to do the out-of-sequence exercise. In addition, appendices and help pages for the different software programs can be found in this book. These may be even more important for reorganizing the exercises.

To accommodate courses that do not have lab periods or that have limited lab time, some exercises have steps that are optional. The exercises are more effective with the inclusion of these steps, but the exercises can be completed and the learning goals reached if these steps are omitted.

The intention of the heavy writing load, as it may appear to many teachers of Astronomy 101, is to appeal to students who are nonscience majors. We who teach this course are typically significantly less comfortable grading a paper than, say, most professors in the English Department, but students from these majors are far more comfortable expressing themselves using prose than mathematics. In an attempt to encourage these students to engage in the process of science, most of these exercises include an optional summary paper (which looks increasingly like a typical lab report as students progress through the exercises). It is recommended that these steps be included, rather than omitted, for the sake of most Astronomy 101 students.

It may seem daunting to grade all these papers, especially if you have a large class, but allowing the process of writing creates a significant gain in the learning process, even if the papers are scored as either complete or not. For many students, the process of writing the paper is where much of the synthesizing of what has happened occurs. It is true, however, that many of these types of students will expect a lot of feedback on their writing.

If you can find a way to provide significant feedback on the papers, you will reap huge benefits of learning in these students. You will engage a population of students who normally are not engaged in courses such as Astronomy 101. The main benefit of this engagement is not new astronomy or physics majors—that is not the goal of Astronomy 101. The benefit will be a scientifically literate citizenry, elected officials, and government employees. This result benefits all of us who are scientists because it promotes science and supports an understanding of the endeavors of basic research.

Earth's Moon in the last quarter phase.

1 Moon Phases and Scientific Models

DESCRIPTION

Scientists create models to attempt to verify their hypotheses and to assist in clarifying theories. A scientific model is a representation of a physical system that may not be entirely accurate but may emphasize a known feature. For example, a model of the solar system may be accurate in the relative sizes of the planets but not in their relative distances. Or, such a model may only be accurate in the order of the planets and not accurate at all in size or distance.

This exercise explores scientific models, hypotheses, and theories. It does this in the context of the Moon's phases.

INTRODUCTION

A hypothesis is a testable explanation of a physical phenomenon. It is not a guess as to why something happens. A hypothesis must be testable; that is, it must be a statement that can be falsified or upheld by an experiment. So, for example, "in the Northern Hemisphere days are longer in the summer and shorter in the winter because Earth's rotational axis is tilted with respect to the plane in which it orbits the Sun, causing the position of the Sun in the sky to appear to change over the course of a year" is a testable statement, which is a hypothesis. This statement is testable because it describes the reason for the phenomenon as the orientation of the Earth's axis with respect to the plane of its orbit around the Sun. This is a property that is observable, so the statement is testable. It sounds like an explanation, but until someone checks to see whether Earth's axis of rotation is tilted with respect to the plane of its orbit around the Sun, it is a hypothesis.

The first part of this exercise is to come up with a hypothesis that explains the Moon's phases. You may have noticed that the Moon appears to change shape over time: sometimes it looks like a crescent, sometimes it looks full (round). Why does this happen?

1. **Write as complete an explanation for this phenomenon as you are able.** (Don't worry about whether your explanation is scientifically accurate at this time. For the purpose of this exercise it is not necessary that your first explanation be correct.) Use the Notes section of this book to record your work during the exercise, then use these notes to produce your final paper.
2. **Now, formulate a statement that constitutes a testable hypothesis.** (A hypothesis that is testable is a statement that predicts how a phenomenon occurs. For example, "the Sun appears to rise in the East because Earth is rotating on its axis in a counterclockwise fashion as viewed from above the North Pole" is a testable hypothesis because one can determine the sense of rotation of Earth

to determine whether this occurs and then create a model to demonstrate how this causes the Sun to appear to rise in the East.) Your hypothesis may be all or part of your answer to question 1, depending on how you wrote your answer to that question.

3. **Now, create a model based on your explanation (answer to part 1) that will allow you to test your hypothesis (answer to part 2).** Use any relevant information you know about the Moon and Earth and how they interact with one another, but don't look up any real scientific models in a book or on the Internet.
4. **Use your model to make a prediction.** You will be observing the Moon every night for a period of time. Make a prediction using your model regarding the appearance (shape) and location of the Moon in the sky at a given time and how these will change (or not) from night to night.
5. **Next, you will observe the Moon every night that it is clear outside (not cloudy) for about three weeks.** Try to situate yourself in the same physical location each time you make your observation. Record your observation as completely as possible. You may use a digital image (from a camera) or a drawing or both. Record the time that you make your observation (it would be a good idea to observe the Moon at about the same time every evening), be sure to note the direction in which you are looking, and the position of the Moon relative to some landmarks, as well as the shape of the Moon.
6. **Once you have gathered all your data, compare these data to your model and hypothesis. Is your hypothesis supported by the data?** Be sure to take into account the position of the Moon and the time you observe it. Remember that the position of an object in the sky at a certain time has something to do with its location relative to Earth and the Sun. That is, if you see the Moon high in the sky at midnight, it is located on the opposite side of Earth from the Sun.
7. **If your hypothesis is not supported by your data, formulate a hypothesis that is supported by your data.** Discuss your results with your instructor.
8. **The final product of this exercise should be a paper that recounts this entire experience, from the first explanation you gave for the Moon phases, to a discussion about how your data support or contradict your hypothesis, to a final explanation of the Moon phases.**

NAME ______________________

DATE ______________________

INSTRUCTOR ______________________

SECTION ______________________

NOTES FOR Moon Phases and Scientific Models

1. Your explanation for the Moon phase phenomenon.

2. Your testable hypothesis statement.

3. Your model based on your explanation.

4. Your predictions for your observations using your model.

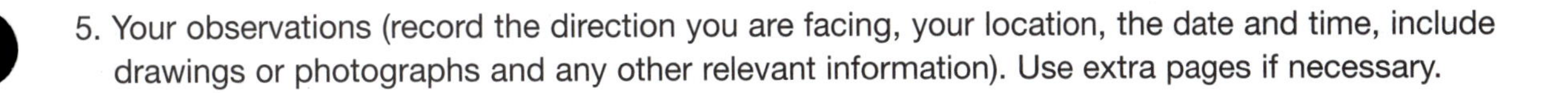

5. Your observations (record the direction you are facing, your location, the date and time, include drawings or photographs and any other relevant information). Use extra pages if necessary.

6. Is your hypothesis supported by your data? If so, demonstrate this by drawing connections between your predictions and your observations. If not, demonstrate this by noting discrepancies between your predictions and your observations and see your instructor.

A portion of the night sky in the northern hemisphere during the early evening hours in the winter and spring.

2 Motions in the Night Sky and the Celestial Sphere

DESCRIPTION

The celestial sphere is a scientific model of the universe. It is not representative of the entire universe, rather it mimics the motions of objects in the night (and day) sky. The celestial sphere model cannot be used to accurately explain the motions of objects in the sky because it is not an entirely accurate representation of the physical universe. A planetarium or planetarium software (such as Starry Night College or the WorldWide Telescope) models of the universe are not entirely accurate representations of the physical universe in the same way that a celestial sphere is not an entirely accurate representation of the physical universe.

This exercise incorporates real observations and a representation (model) of the universe (a celestial sphere, a planetarium, or planetarium software) and examines the usefulness of these models while identifying the model's shortcomings.

INTRODUCTION

A scientific model is one that is based on scientific observations and represents a physical system that accurately reflects at least one aspect of the physical system. For example, a scientific model of the solar system may include an accurate representation of the planets' sizes relative to one another, but not their distances from one another. Such a model would still be considered a scientific model. Several scientific models have been developed that mimic, with some accuracy, the motions of objects in the sky. In this exercise, you will observe the motions of the stars in your night sky and compare your observations to one of the several models astronomers have developed to mimic these motions. Then, you will critically analyze the usefulness of the model to which you compared your observations.

First, you will make some observations. You already know that the Sun rises in the morning and sets in the evening, bringing on night. At night we can observe stars, the Moon, and some planets without the aid of a telescope or binoculars. You will need to find a location that is dark enough that you can see at least ten stars. It would be best if the stars you can see are in a memorable pattern or part of a constellation, but depending on your proximity to a large city, you may or may not be able to find a spot from which you can see the Big Dipper, for example.

PART I: OBSERVATIONS OF THE NIGHT SKY

1. **Before you make any observations, write down a prediction. Considering that the Sun and Moon both rise and set, predict what the motions of the stars you observe will be.** Try to be as

complete as possible with your prediction. Identify the direction of the motion you predict as well as the rate of motion. Use the Notes pages associated with this exercise to record your prediction and observations. Then, for your final paper, these notes can be used as an outline.

2. **Make and record an observation of the night sky.** If you are in the Northern Hemisphere, face any direction except north. Observe the night sky and make a drawing representing what you observe. Include some of your horizon as well as at least ten stars or patterns of stars or other celestial objects you see in the night sky. It's not important that you can identify the constellations or stars, but you may try to do that if you wish. Note the time of your observation and include cardinal directions (north, south, east, or west) on your drawing.

3. **Make and record a second observation of the night sky.** Once you have finished your drawing of your first observation, stop and do something else. After about 1 hour, come back to the same location, facing the same direction, and make a new drawing of what you see. Try to find at least some of the same stars or patterns as before. Again, be sure to include some of your horizon as well as the stars and other celestial objects you see in the night sky. Note the time of your observation, and include cardinal directions (north, south, east, or west) on your drawing. (It is not necessary, but you could continue to make observations 1 hour apart for as long as you are able to get more accurate data.)

4. **Compare the two observations you made.** Knowing the time between the observations, compare the two and determine whether you observed motion. Then determine the direction and rate of the motion you observed. (About how far did the stars move and in what general direction?) To know whether you've analyzed your data sufficiently, know the time it would take for a star you observed to set (disappear beyond the horizon).

PART II: USING AN ASTRONOMICAL MODEL

5. **Use an astronomical model to check your observations.** Set up a celestial sphere, planetarium, or planetarium software to show the night sky at the time and day that you made your first observation. Now you can check to determine what objects you saw. Identify at least ten objects (stars or planets or other celestial objects) from your drawing using the model you chose. Determine the time it would take for a star you observed to set (disappear beyond the horizon).

   a. **Celestial Sphere.** Using a celestial sphere, you will need to move the Sun to the correct position for the day that you made your observation, then rotate the sphere to the time that you made your observation. You can then check your observation by fixing the horizon to match your latitude. On the sphere, larger dots are brighter stars. You will not be able to identify planets or the Moon with this tool.

   b. **Planetarium.** The planetarium can be set to mimic exactly the time and day that your observation was made. The horizon in the planetarium will be more extensive than that which you observed. Brighter objects will appear larger in the planetarium "sky." Most planetaria include constellation outlines and a moon. You may need to use a star chart to determine the names of the stars you observed.

   c. **Planetarium Software.** Stellarium, Starry Night College, WorldWide Telescope, and SkyGazer all may be set up to show the night sky at any time and for any location on Earth. Also, each of these software programs will label constellations and allow the user to determine the names of any objects, including galaxies, nebulae, planets, and stars.

6. **Compare your observations to your prediction.** Answer the following questions regarding your observations and prediction. Each answer should be at least a few sentences, not just a single word or sentence. Try to give as complete an answer as possible.

   a. **Did the stars move as you expected?** (If you did not expect them to move, rephrase this question to read, Did the stars stay stationary as you expected?) If the answer is no, describe in detail how the motion of the stars differed from your prediction.

b. **Compare the motions of the stars that you observed to the motion of the Sun.** Use the planetarium software, the celestial sphere, or a planetarium to simulate the motion of the Sun.

c. **Describe in detail, using only your observations and the model you have (celestial sphere, planetarium, or planetarium software), the reason for the apparent motions of the stars across the sky.** It is not important that this answer is scientifically correct. It is more important that you base your explanation on the model you are using and the observations you made. (Are the stars moving around Earth, or is Earth's motion causing the stars to appear to move? What does the model you are using indicate and why?)

d. **Describe, in detail, using resources in your textbook, in the library, or on the Internet, the reason for the apparent motion of the Sun across the sky.** For this response, be sure to cite the resources you use. This answer should be a scientifically correct explanation for the apparent motion of the Sun across the sky. Check with your instructor to make sure you have the correct scientific explanation before you proceed to the next question. Would this reason explain the star motions you observed?

PART III: EVALUATION AND FINAL PRODUCT

7. **Evaluate the model by answering the questions below.**

   a. **Compare the model you have (celestial sphere, planetarium, or planetarium software) to the correct scientific explanation.** Does the model demonstrate a correct explanation for the motions of the Sun and stars?

   b. **Evaluate the usefulness of the model for demonstrating the motions of objects in the sky.** Does the model replicate the motions accurately? Is the model easy to use?

   c. **Evaluate the usefulness of the model for explaining the reasons for the apparent motions of objects in the sky.** Does the model create any misconceptions about the motions of objects in the sky?

OPTIONAL Step 8. Your final product should be a paper that describes the experiment (what you set out to understand and how you set out to understand it), the hypothesis (your prediction), the data collection (how you made your observations), the data (your observations), the analysis (how you compared your data to a model and your answers to questions 6a-d), and your conclusions (your answers to the questions posed in step 7).

NAME ______________________

DATE ______________________

INSTRUCTOR ______________________

SECTION ______________________

NOTES FOR Motions in the Night Sky and the Celestial Sphere

1. Write your prediction for the motion of the Sun and stars. Make sure to identify direction and rate of motion.

2. Record your first observation of the night sky here.

3. Record your second observation of the night sky here.

4. Compare your two observations. Note any changes in position of objects. Write down any notes you have on what you have observed, here.

5. Write down any notes you have regarding the observations you make with the model (celestial sphere, planetarium, or planetarium software).

6. a. Did the stars in the model move as you predicted? If not, describe in detail how the motion (or lack thereof) differed from your predictions.

   b. Compare the motion of the stars with the motion of the Sun.

c. Describe, in detail, using only your observations and the model you have, the reason for the apparent motion of the stars across the sky. Are the stars moving around Earth, or is Earth's motion causing the stars to appear to move? What does the model you are using indicate and why?

d. Describe, in detail, using resources in your textbook, the library, or the Internet, the reason for the apparent motion of the Sun across the sky.

7. a. Compare the model you have to the correct scientific explanation. Does the model demonstrate a correct explanation for the motions of the Sun and stars? How does it do so, or how does it fail?

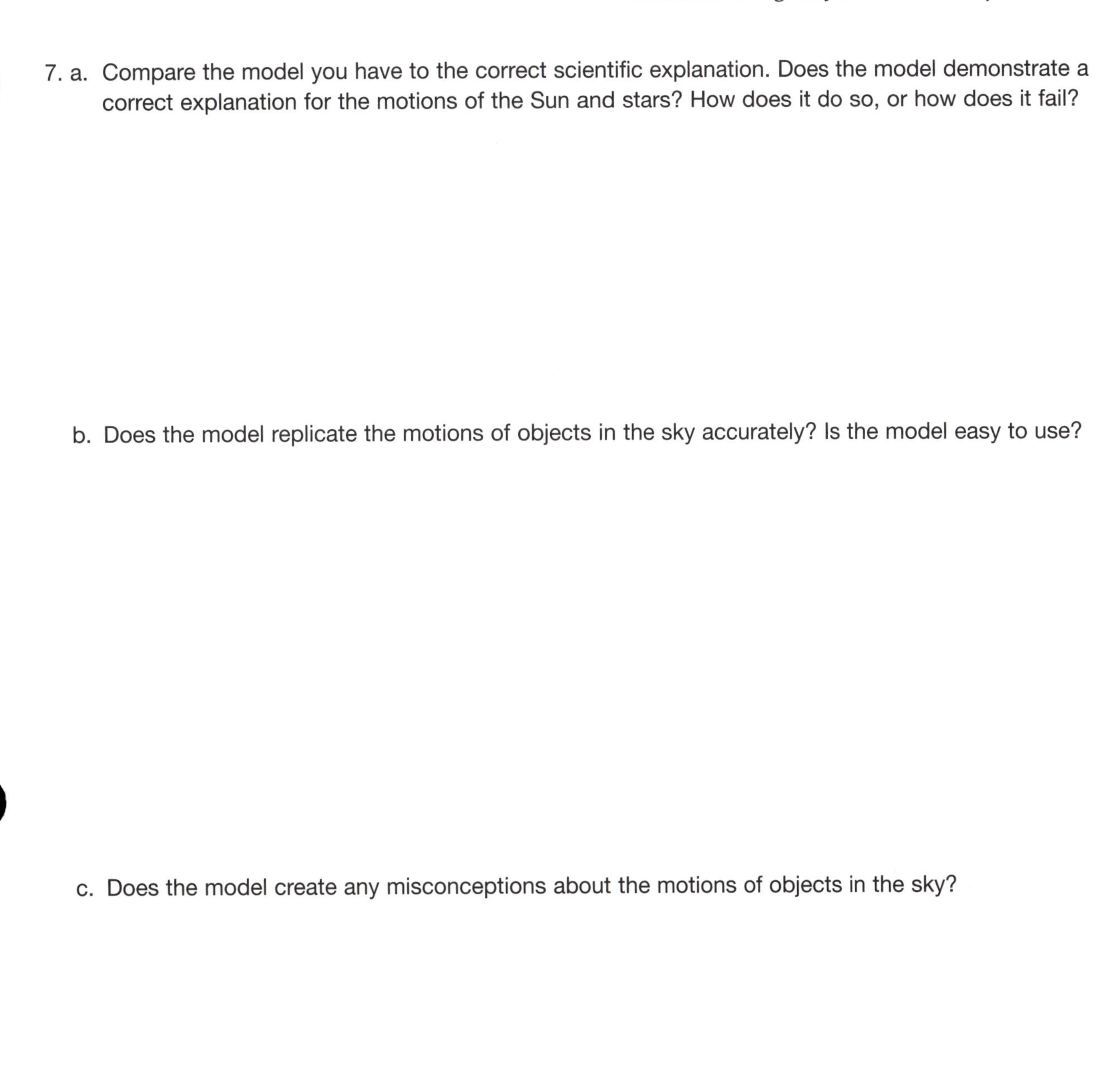

   b. Does the model replicate the motions of objects in the sky accurately? Is the model easy to use?

   c. Does the model create any misconceptions about the motions of objects in the sky?

The pattern of the Sun's positions shown here is known as an analemma.

3
The Sun Through the Seasons

DESCRIPTION

At most locations on Earth, there are seasonal changes throughout the year. Because one year is the length of time it takes Earth to orbit the Sun, these changes must be associated in some way with Earth's journey around the Sun. In addition to the seasonal weather changes, there are seasonal changes that involve astronomical observations. One change is in the pattern of stars observable at night, which changes throughout the year. Another change is in the location of sunrise and sunset, as well as the path of the Sun across the sky throughout the year. Observations of these phenomena can reveal the cause of the changes in the weather patterns. Although the change in location on the horizon of sunrise and sunset does not cause the weather patterns changes per se, the changes we observe in this phenomenon are due to the same characteristic of Earth that causes the seasonal changes.

This exercise incorporates planetarium software observations to create a model of the Earth–Sun system and deduce the cause of the seasons.

INTRODUCTION

In this exercise, you will develop a scientific model of the Earth–Sun system based on the observations you make. You will start with the knowledge that Earth orbits the Sun and that the time it takes Earth to complete a full orbit is exactly one year. Your model will incorporate this information as well as what you learn from your observations.

1. **Hypothesize the cause of the seasonal changes observed on Earth**. It does not matter whether your hypothesis is correct or not. What matters is whether you can be complete enough about your hypothesis to predict the outcomes of your observations. It is also important that your hypothesis be testable. After you have written one down, ask yourself whether you could verify this hypothesis with data you could collect. Use the Notes at the end of this exercise to record your hypothesis and observations. When you write your final paper, these notes will serve as an outline.

2. **Predict the outcomes of the following observations, assuming your hypothesis is correct**. You will observe the position of the sunrises and sunsets from your location on Earth at least two times a month for one year (using planetarium software, so you can do this quickly). You will also measure the length of time it takes the Sun to cross the sky at least two times a month for one year (again, using planetarium software). Finally, you will note the name of the constellation closest to the cardinal point east at sunset at least two times per month for one year (using planetarium software).

3. **Make the observations.** Set up your planetarium software to begin on the first day of this year (January 1). For the purpose of this lab, sunrise is when the full disk of the Sun is just visible above the eastern horizon and sunset is when the full disk of the Sun is just visible above the western horizon. Measure the position of the Sun (relative to east) at sunrise on this day. Note the time of sunrise. Fast-forward the software to sunset. Measure the position of the Sun (relative to west) at sunset on this day. Note the time of sunset. Look east at sunset and note the name of the constellation closest to east at this time (even if the full constellation is not visible). Change the date to January 15 and repeat. Continue to gather data every two weeks through December 31 of this year. Make sure you have at least 24 data points for each phenomenon you are observing.
4. **Describe your observations and compare them to your predictions**. Look for patterns in the observations you have made, and describe what you see. Did your observations support your hypothesis? If not, can you formulate a new hypothesis based on these observations?
5. **Discuss your results with your instructor or other class members**. If your hypothesis was incorrect, what could your data mean? If you had to develop a new hypothesis, is there some observation you could make to confirm that your new model is correct?

OPTIONAL Part 6. Devise an experiment that demonstrates how the directness of light rays could be related to the seasonal weather changes on Earth. Use a flashlight and a globe or other representation of Earth to demonstrate how the surface of Earth could be warmed by the Sun differently during the different seasons.

OPTIONAL Part 7. Write a final paper recounting the experiment. Show your original hypothesis, your observations, analysis, and final model.

NAME ______________________________

DATE ______________________________

INSTRUCTOR ______________________________

SECTION ______________________________

NOTES FOR Sun Through the Seasons

1. Use this space to write out your hypothesis/explanation for the changing seasons observed on Earth throughout the year.

2. Use this space to predict the outcomes of the following observations:

   a. Position of sunrise/sunset two times per month over a one-year period.

   b. Length of time it takes for the Sun to go from sunrise to sunset two times per month over a one-year period.

   c. The name of the constellation closest to east at sunset two times per month over a one-year period.

3. Use this space to record your observations:

Date	Sunrise position (relative to east)	Sunset position (relative to west)	Length of day (hours)	Constellation in east at sunset
January				
January				
February				
February				
March				
March				
April				
April				
May				
May				
June				
June				
July				
July				
August				
August				
September				
September				
October				
October				
November				
November				
December				
December				

4. Use this space to summarize your observations. Describe any trends or similarities you notice in your data.

5. Use this space to compare your observations to your predictions. Note all similarities and differences. Consider the implications of any differences on your hypothesis.

6. Reconcile your observations with your hypothesis. If your data are not as you predict, what explanation for the cause of the seasons is consistent with your data? Is there more than one possible explanation? What other data could you gather to confirm this new explanation and rule out any others?

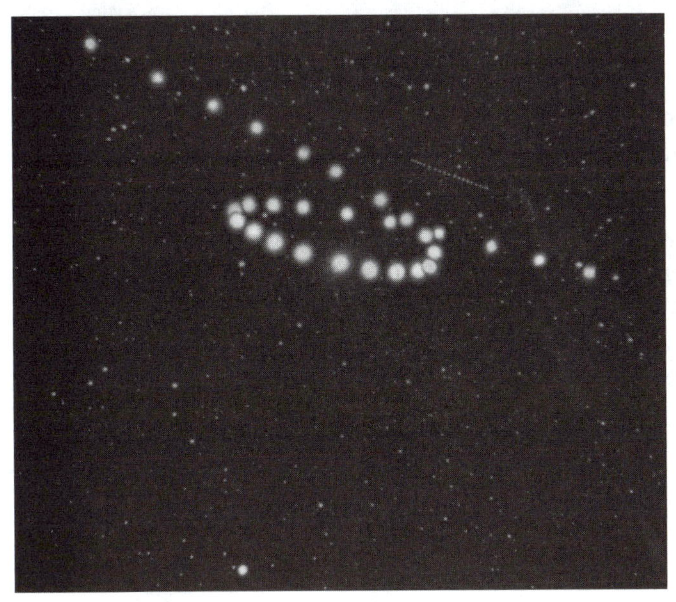

A pattern of Mars's positions against the stars.

4 Mars's Motion and Models of the Solar System

DESCRIPTION

Early models of the solar system were simple and did not incorporate all observed phenomena. Aristotle had all the planets and the Sun orbiting in circular orbits around Earth. To incorporate observations of the outer planets exhibiting retrograde motions, Ptolemy's model included epicycles, which almost matched the observations. Until the Renaissance, when people were more open to accepting new ideas, and until observations of the planets were accurate enough to rule Ptolemy's model out, a more accurate model of the solar system was not accepted.

This exercise explores data-driven models in the context of Mars's retrograde motion.

INTRODUCTION

Developing an accurate model of the solar system is like trying to map a city standing in one location. If the location is just right (high enough and central enough), you can see what you need in order to map things out, but if you're not in the right place and not able to observe in all directions, you could miss something very important. Keep in mind the fact that the solar system is huge. If you put it on the scale of a football field, Earth would be the size of a peppercorn and the Moon would be about 1 cm away from it. Humans have never been farther than the Moon. Only a few unmanned satellites have been to the planets at the farthest reaches of our solar system. Even so, centuries ago, humans were able to deduce some complex aspects of our solar system simply by watching how the planets move against the background of stars over a period of time.

The first part of this exercise involves gathering some data on the motion of Mars relative to the background stars.

1. **Set up your planetarium software to observe Mars.** Watch the motion of Mars relative to the stars by setting the time step you observe to "days" and centering your view on Mars. Make sure you observe the motion of Mars carefully for at least 2 years of "real" time. Describe the motion you observe. Share your description with at least two other students and your instructor to make sure you have observed the phenomenon of retrograde motion. Note the dates when Mars is in retrograde motion on the Notes page found at the end of this exercise.
2. **Write out an explanation for this "backward" motion of Mars.** It is not important that your explanation is scientifically correct. However, it is important that you give a complete explanation that can be tested. Check your explanation with your instructor to make sure it is complete.

OPTIONAL EXTENSION. Observe the motions of Jupiter, Saturn, and Venus to see whether they exhibit the same motion. How do these observations modify or support your explanation for Mars's retrograde motion?

3. **Observe the actual motions of Earth and Mars.** View the live solar system by going to http://ftp.fourmilab.ch/cgi-bin/Solar and entering the date (under UTC) closest to the time you found Mars doing its backward motion. Press the **Update** button. Note the positions of Mars and Earth. Then, enter a date that is about one month later. Notice the change in the positions of Mars and Earth. Continue to change the date by one month at a time until you can see how Mars and Earth are moving during the time that Mars is going backward.
4. **Record positions and dates so that you can map the apparent motion of Mars along with the relative positions of Mars and Earth in their paths around the Sun.** Note the date and time of each observation (both with the planetarium software you are using and with the Live Solar System page). Use these data to demonstrate the cause for the apparent motion of Mars.

OPTIONAL STEP 5. Write a paper describing this entire exercise, starting with your original hypothesis. Include the data you've gathered as well as any other relevant information or images you would like to add.

OPTIONAL EXTENSION. Find out about Ptolemy's model of the universe and figure out how his model explained Mars's retrograde motion.

NAME ______________________

DATE ______________________

INSTRUCTOR ______________________

SECTION ______________________

NOTES FOR Mars's Motion and Models of the Solar System

1. Describe the motion of Mars that you observe. In which direction, relative to the stars, does Mars move? Describe any anomalies in Mars's motion that you observe.

2. Using what you know about the solar system, try to explain why Mars appears to move the way it does, relative to the stars.

3. Use this space to record the positions and dates for Mars and Earth that you need in order to show why their motions in the solar system and Mars's motion in our night sky are the way they are.

4. Using the information you recorded above, explain why Mars appears to move the way it does in our night sky.

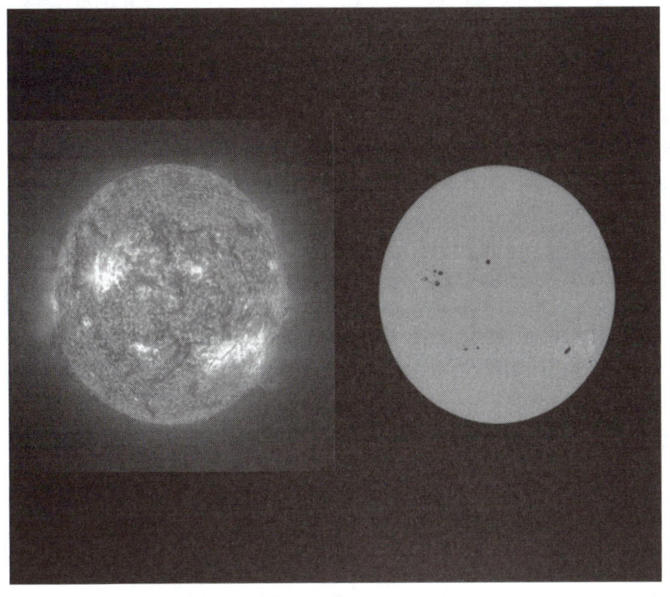

The Sun shown through two different filters.

5
How Did Galileo Go Blind? Or, It's the Data, Stupid!

DESCRIPTION

Four hundred years ago there was a man who wanted to understand the universe and who found that what humans knew was limited. The brightest minds had come up with ideas of how things worked that simply contradicted what he knew to be true. As a result, he began to do things that no one expected. First, he dropped objects from the top of the Tower of Pisa (not leaning yet). Then he took a small telescope that he had made based on a design he found in a publication by a Dutch optician, and he pointed it to the heavens. He looked at the Sun, the Moon, Venus, Jupiter, Saturn, and the Milky Way. What he observed challenged everything the smartest people were telling him must be true. Nonetheless, he carefully made observations and recorded his data. He developed models based on these data that were controversial because they were not consistent with the dogma of the time. Today, we know that many of Galileo's ideas were correct and that his observations marked the beginning of modern experimental science.

This exercise explores modern observations of the Sun, using data (observations) to explain observed phenomena (such as sunspots, coronal mass ejections, and solar flares).

INTRODUCTION

In science, a model or a description without data to support it is as accurate as a fairy tale. Theories in science rely on strong support from many different data sources. Galileo's contribution to science was first and foremost that he supported everything with actual data. He was one of the first in a long time to do this. To understand the Sun (the object Galileo observed that likely caused his blindness), we must gather together many observations of this object and try to understand them.

This exercise will incorporate data from many sources on the Internet to help us understand some of the phenomena of solar activity, namely, sunspots, flares, and coronal mass ejections. Our working hypothesis is that sunspots, flares, and coronal mass ejections are caused by distortions in the Sun's magnetic field. We will now look at the data we find and see if they support this hypothesis.

1. **Go to the following URL and look for data that could support or refute the above hypothesis statement:**
   http://sohowww.nascom.nasa.gov/home.html. Click on the link titled "The Sun Now." Go through the data and find images that show sunspots (continuum images), flares (EIT 171, 195, or 284 images), coronal mass ejections (LASCO C2 and C3 images), as well as magnetograms. Peruse the site a little and get familiar with where these images are found and how to view them.

2. **Identify the observed features and compare their locations.** Go to Search and Download Data; if there are no sunspots on the image you view, verify that there are also no flares or coronal mass ejections (CMEs) or magnetic disturbances, then look for data on a day when there were features to observe. Note that, on the magnetogram, dark and light spots indicate areas of anomalously strong magnetic polarity. Keep track of your observations. For each date, record any sunspots, CMEs, and evidence of strong magnetic polarity.

OPTIONAL Part 3. Using the images available, identify and describe the following features: sunspot, prominence, and flare. Answer the following questions: In which images are sunspots most easily seen? Why? Knowing that different layers of the Sun's atmosphere are observed with different images and that each layer is associated with a particular temperature, explain why sunspots are dark.

4. **Do the data you've found support the hypothesis that these phenomena are all related to magnetic anomalies?** Examine the data you recorded and look for correlations. For how many instances of a sunspot is there a magnetic anomaly? For how many instances of a sunspot is there a flare? Are coronal mass ejections only associated with sunspots? You can quantify your correlations by looking at the percentage of your observations that follow these trends. Discuss whether these correlations support the hypothesis.

5. **Use the *SOHO* Movie Theatre to create a movie and observe the motion of sunspots on the Sun. (If this is a relatively inactive time, use October 15, 2003 through November 15, 2003.)** For the time period you use, watch corresponding movies of EIT 284, LASCO2, and the magnetogram. Compare the phenomena of sunspots, CMEs, and the magnetogram in these movies. The accepted scientific models of solar activity show the Sun's magnetic field, which is embedded in the material that makes up the Sun, winding up to cause the observed magnetic anomalies. How well does the magnetogram movie support this assertion?

OPTIONAL Part 6. Write your results in a report. Describe the data you used in this experiment and include any images you can (be sure to name the source for each image). Explain your method of analysis and your conclusions.

OPTIONAL Extension Part 7. Explore the NASA Web sites to learn more about how the magnetic field causes sunspots, flares, and coronal mass ejections. Learn how these phenomena affect Earth (i.e., aurora borealis, radio static). Include this information in your final paper.

NAME______________________________

DATE______________________________

INSTRUCTOR______________________________

SECTION______________________________

NOTES FOR How Did Galileo Go Blind? Or, It's the Data, Stupid!

1. Use this space to record your observations. Make sure to note the date and type of image observed along with the presence or not of sunspots, coronal mass ejections, or strong magnetic polarity.

2. Analyze the data you collected, quantifying, where possible, your correlations. Discuss whether these correlations support the hypothesis.

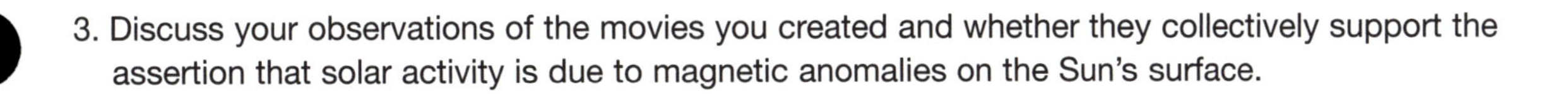

3. Discuss your observations of the movies you created and whether they collectively support the assertion that solar activity is due to magnetic anomalies on the Sun's surface.

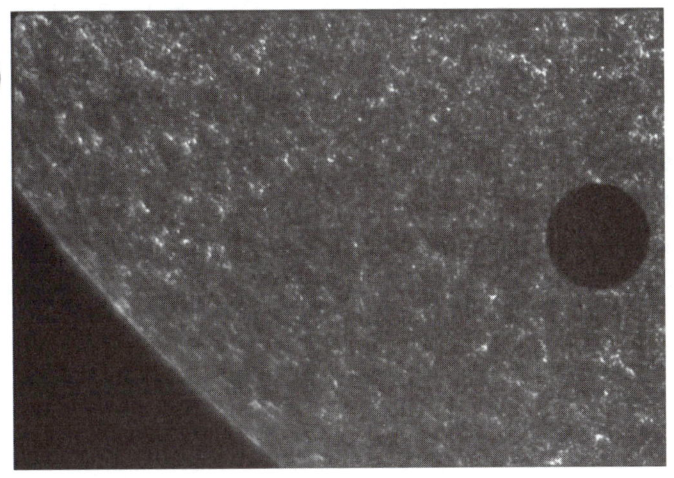

Venus against the disk of the Sun. As seen from Earth, this is a rare event. It occurs twice (eight years apart) in about a century and a quarter.

6 What's an Astronomical Unit?

DESCRIPTION

The average distance between the Sun and Earth is called an astronomical unit (AU). Before astronomers knew a precise value for the astronomical unit, they were able to determine the distances to other solar system objects using the astronomical unit as a scale. It wasn't until the rare event of the transit of Venus across the Sun that astronomers were able to make an accurate measure of the astronomical unit. This is a case of astronomers devising a clever experiment to make a measurement.

This exercise gives you the necessary tools to determine and make measurements (during a transit of Venus) needed to calculate the length of an astronomical unit. In the process of this exercise, you will use planetarium software to observe the recent transit of Venus and make measurements you will use to calculate the length of an astronomical unit.

INTRODUCTION

So, what is the length of an astronomical unit, and how did astronomers measure this distance? The actual length of an astronomical unit is 1.496×10^{11} m, which one can look up in any astronomy textbook or online. The more interesting question is: How did astronomers measure this distance? Actually, several measurements were made over a period of at least 250 years. The measurements made most recently, as one might expect, are more accurate, however. Each measurement of the astronomical unit uses the same principle: parallax.

Parallax is a phenomenon that is most readily observable by holding an object (such as a pen or a pencil) in front of your face at arm's distance. When you close one eye (say, your left eye) and observe the object with your right eye, then switch and close your right eye and open your left eye, the object appears to move with respect to the background. This is the parallax effect. Nearby objects appear to move relative to distant objects when observed from different vantage points.

To measure the length of an astronomical unit, one has to observe some phenomenon that occurs within the solar system from two different locations on Earth. The event must involve a nearby object and a more distant object for which the relative distance between the two objects is known in astronomical units. The first accurate measurement of the length of the astronomical unit was made during the transit of Venus that occurred in 1769. In 2004, Venus transited the Sun yet again. It will do so yet one more time in 2012.

1. **Draw a diagram showing how the appearance of Venus (a silhouette) relative to the surface of the Sun will be different viewed from two different places on Earth.** The scale of the objects in your drawing is not as important as your ability to demonstrate how parallax will allow you to determine the distance of an astronomical unit. Label the distance between Earth and Venus $R_E - R_V$.

Label the distance between the Sun and Venus R_V and the distance between Earth and the Sun R_E. Label the angle between the two positions on Earth θ. The angle between the two positions on the Sun will be the same, because when two lines intersect, the opposite angles are equal. Use the Notes space found at the end of this exercise to record your responses to these steps. These notes can be used as an outline when you are preparing your final paper for submission.

2. **Use geometry to determine the distance between Earth and Venus.**[1] If you split the triangle formed by the two positions on Earth and the position of Venus in half, you have two right triangles. One side of one of the right triangles is the distance between Earth and Venus, and one side is half the distance between the two positions on Earth. The angle opposite the side that is half the distance between the two positions on Earth is half the angle identified as θ. The tangent of an acute angle in a right triangle is equal to the length of the side opposite the angle divided by the length of the side adjacent to the angle. Show how the distance between Earth and Venus could be determined knowing the angle and the distance between the two points on Earth. Check this part with your instructor before continuing.

3. **Use relative distances to show how the length of the astronomical unit can be determined knowing the distance between Earth and Venus.** Astronomers knew that if the distance between Earth and the Sun is 1, the distance between Venus and the Sun is 0.7.[2] This means that the distance between Earth and Venus is $1 - 0.7 = 0.3 = R_E - R_V$. Using what you have developed for step 2, show how you can now determine the length of the astronomical unit.

OPTIONAL Step 4. Determine the measurements needed during a transit of Venus from two different locations. Identify two locations that seem appropriate for this experiment. Be sure to identify the necessary information for each observation. Do you need to know the time of the observation? Do you need to know the position of Venus? Relative to what? Check this part with your instructor before continuing.

ALTERNATE OPTIONAL Step 4. Choose a single time during the middle of Venus's transit across the Sun. Figure out which longitude on Earth is experiencing noon at that time. Change the latitude as far south as possible until the Sun is sitting on the horizon. That will be a point on the "bottom" of the Earth. Now, change the latitude until the Sun is exactly overhead. That will be the "middle" of the Earth. Finally, change the latitude as far north as possible until the Sun is exactly on the horizon. (Note: you will have to increase the longitude by 180 degrees.) This will provide a point on the "top" of the Earth. Venus will not be in front of the Sun from all three locations, but it should be in front of the Sun from two of them. Use the distance between the two points observed when Venus is in front of the Sun. Top and Middle are separated by 1 Earth radius; Bottom and Middle are separated by 1 Earth radius; and Top and Bottom are separated by 2 Earth radii.

5. **Use planetarium software to make the necessary measurements.** Record the information you think is important to make the calculations you need to make. The planetarium software should make it easy to get the information needed.

6. **Calculate the length of the astronomical unit.** Once you have done this, compare your value to the accepted value of 1.496×10^{11} m. How close did your measurement come to the accepted value? What is the percent difference? Why is your value different? What assumptions did you make in your method of calculation that could have contributed to this error? Could you have made more accurate measurements? How could you have improved the experiment?

7. **Do the experiment one more time for two different locations using the most accurate measurements you can get from the planetarium software you have.** Calculate the length of the astronomical unit again. Is your second answer more accurate? Why or why not?

OPTIONAL Step 8. Your final product should be a paper explaining the observations you made and describing the calculations you did to determine the value of the astronomical unit.

---

[1] For review of geometric principles that can be used for this calculation, see the appendix.

[2] The Titius-Bode Law, derived from the works of Johann Daniel Titius (1766) and Johann Elert Bode (1768), is the source of this knowledge. This law predicts the positions of Mercury, Venus, Earth, Mars, Jupiter, Saturn, and Uranus, as well as the location of the asteroid belt.

NAME ______________________

DATE ______________________

INSTRUCTOR ______________________

SECTION ______________________

NOTES FOR What's an Astronomical Unit?

1. Draw a diagram showing how the appearance of Venus (a silhouette) relative to the surface of the Sun will be different viewed from two different places on Earth.

2. Use this space to work out the distance between Earth and Venus by using the properties of similar triangles and the small angle approximation.

3. Using what you have done in part 2, show how you can now determine the length of the astronomical unit, knowing the distance between Earth and Venus.

4. Use planetarium software to make the necessary measurements and record the information you think is important to make the calculations needed.

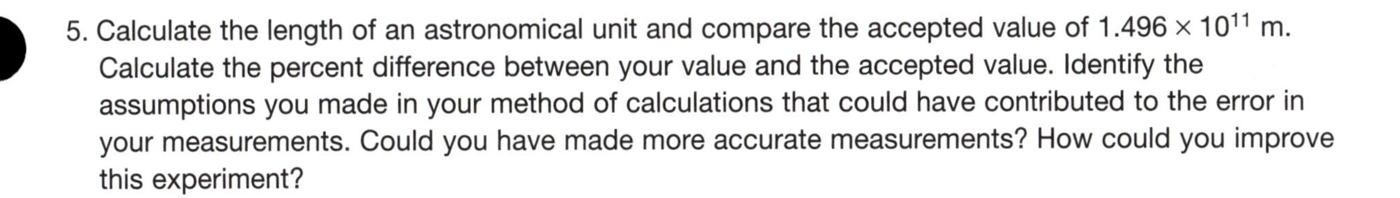

5. Calculate the length of an astronomical unit and compare the accepted value of 1.496×10^{11} m. Calculate the percent difference between your value and the accepted value. Identify the assumptions you made in your method of calculations that could have contributed to the error in your measurements. Could you have made more accurate measurements? How could you improve this experiment?

6. Use this space to record your second set of measurements and calculations. Compare the results of the two experiments. Which is more accurate? What may have caused the inaccuracies in your results?

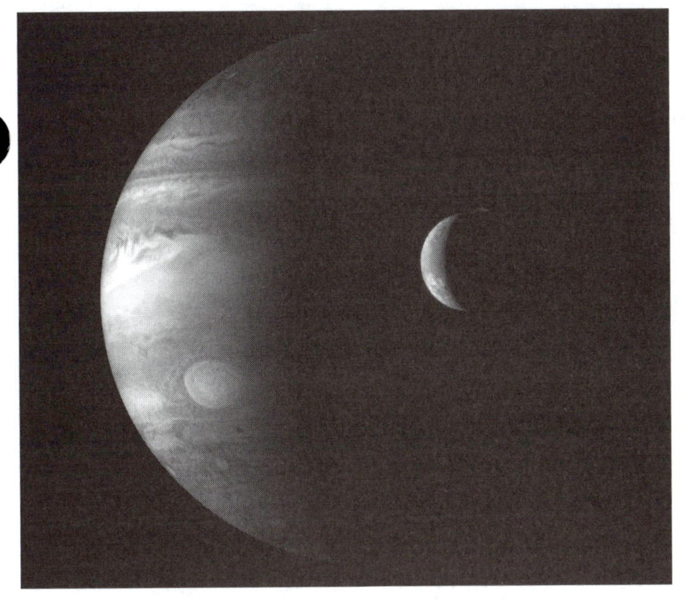
Jupiter and its moon Io.

7 How Do We Know the Mass of Jupiter, Anyway?

DESCRIPTION

One of the most frustrating and simultaneously fascinating things about astronomy is that, for the most part, none of it is tangible. That is to say, we cannot just go to the Sun and measure its temperature, or take a sample of a planetary nebula to see what it's made of, or put Jupiter on a scale and weigh it. Everything in astronomy is learned by devising clever experiments to make the necessary measurements. All astronomical data use electromagnetic radiation of some sort. Astronomers, therefore, must understand everything they can about how light can give information about the physical nature of an object. For example, because we cannot go to the Sun and measure its temperature, we study the spectrum of the Sun; from the absorption lines we observe in its spectrum, we can deduce something about its temperature. We can also look at the spectral energy distribution of the Sun's light (look at how much light we receive from the Sun at various wavelengths) and, using what we know about blackbodies, determine the temperature of the surface of the Sun.

This exercise gives you the necessary tools to design a clever experiment to measure the mass of Jupiter. Work in a group to gather the data efficiently, using planetarium software that allows you to simulate the motions of the satellites of Jupiter for any time you choose.

INTRODUCTION

So, how did astronomers measure the mass of Jupiter? The answer is by some clever experiment that incorporated observations only from Earth. In this exercise, you will design and perform a clever experiment to measure Jupiter's mass. This experiment will involve observing the motions of objects and understanding how their motions are related to their masses. To do this, let's review some of the tools of astronomy you will need to use.

By now in your introductory course of astronomy, you probably know Kepler's laws of motion, including the third law, loosely referred to as $P^2 = a^3$. Kepler did an excellent job of describing the motions of the planets that he observed, but he did not explain the reason for the motions. (Kepler's laws are so named because they accurately describe the motion without giving an explanation. In science, a *law* means exactly that: a [usually] mathematical relationship that accurately describes an observed behavior, but does not explain it.)

You have probably been introduced to Newton's version of Kepler's third law, in which Newton explained what Kepler observed using his universal law of gravitation. (Again, the use of the term *law* means that although gravitation explains the motion observed by Kepler, the universal law of gravitation does not include an explanation for why gravity exists or behaves the way it does.)

1. **Explain what Kepler's third law ($P^2 = a^3$) describes.** How could one use Kepler's third law to determine the length of a year on Jupiter, for instance? Conversely, how could it be used to calculate the average distance between the Sun and Jupiter? What are the units of P and a in this equation? Use the Notes pages found at the end of this exercise as a space for recording your responses to these questions.
2. **Newton's version of Kepler's third law of planetary motion can be written as follows:** $P^2 = \frac{4\pi^2 a^2}{G(M + m)}$ **. Define each variable (P, a, G, M, m) and state the units for each.** **Hint: The units are not the same as in Kepler's third law.** If you don't know the units for G, you should be able to figure it out by plugging in the rest of the units. Remember that all the units should be in the same system; that is, if you use kilograms for mass, you need to use seconds and meters for time and distance, respectively.
3. **The form of Kepler's third law above is useful even for systems for which the Sun is not the central mass. Jupiter has several natural satellites that obey this law of motion. How could this equation be used to determine the mass of Jupiter?** What values would you need to know to plug into the equation above to make this calculation? What would a, P, M, and m be? (You don't need to know the actual numbers to answer this, just what each variable represents in terms of this specific problem.)
4. **List the observations and measurements needed to determine the period of rotation of a satellite around a planet.** It is not necessary to know the satellite's orbital period. You may simply determine how many orbits you would need to observe and at how many points during that orbit you would want to take data.
5. **List the observations and measurements needed to determine the average distance (semi-major axis length) between a satellite and the planet it orbits.** Assume you know the distance between the planet and Earth. Use the principles you learned in the "What's an Astronomical Unit?" exercise to determine the distance between the planet and the satellite.

OPTIONAL Step 6. Describe an observational experiment you could do to measure the mass of Jupiter using Newton's version of Kepler's third law to calculate the mass. Be sure to list all the observations you would need to make (how many or how frequently and of what), what kind of data you would need to gather (what measurements), and how the data would be used to calculate the mass of Jupiter. Before proceeding to the next step, check this experiment description with your instructor.

7. **Partner with at least one other classmate and use one of the following planetarium software programs to simulate the motions of the Galilean satellites around Jupiter: Starry Night College, Stellarium, SkyGazer, or the WorldWide Telescope. Make the observations and measurements you need for your experiment.** If you have decided to observe more than one satellite, share the observation duties by having each group member observe a different satellite.
8. **Analyze your data and make graphs, if necessary.** Based on the plan laid out for this exercise (either by you or the instructor, depending on how step 6 was treated), determine the values for the variables in the equation for the mass of Jupiter. You may need to make plots or graphs to determine the values of some quantities. For example, to determine the period, it might be necessary to plot the distance from Jupiter versus time measured for a satellite so you can determine the length of time it took for the satellite to orbit Jupiter.
9. **Calculate the mass of Jupiter using the data you gathered and compare it to the number in your textbook or other resource.**

OPTIONAL Step 10. The final product will be a paper describing the method you used to determine the mass of Jupiter. Explain why your method works, and show all your data and calculations. In the conclusions section of your paper, compare your result to other resources and explain any differences you notice. Be sure to consider all the assumptions you made in designing the experiment when you are trying to account for differences in the masses.

NAME ______________________

DATE ______________________

INSTRUCTOR ______________________

SECTION ______________________

NOTES FOR How Do We Know the Mass of Jupiter, Anyway?

1. Explain how one could use Kepler's third law to determine the length of a year on Jupiter or how it could be used to calculate the average distance between the Sun and Jupiter. Be sure to give the units for P and a in the equation.

2. Define and state the units for each variable in Newton's version of Kepler's third law: $P^2 = \frac{4\pi^2 a^2}{G(M + m)}$.

3. List the observations and measurements needed to determine the period of rotation of a satellite around a planet.

4. List the observations and measurements needed to determine the average distance (semi-major axis length) between a satellite and the planet it orbits.

Spectra of stars. The dark lines are absorption lines due to gases in the atmospheres of the stars. Although stars are all composed of the same materials, due to their varying surface temperatures, some stars have spectra that show the presence of different elements.

8
What Are Stars Made Of?

DESCRIPTION

When astronomers first started to study stars, they noticed that stars appeared to have different intrinsic colors. Later, with more sophisticated instruments, these color differences were verified to be distinct differences in the amount of light emitted at each wavelength, meaning the difference in color was real and not perceived. More recently (in the early 20th century), astronomers used spectrographs to separate the light from stars even further, and they noticed that the stars gave off absorption spectra. Different color stars gave off different spectra. It took decades for astronomers to put together the meaning of all these different pieces of data, but eventually they were able to determine what causes the different spectra and what elements make up the stars.

In this exercise, you will look at stellar spectra and blackbody curves of stars to determine the stars' temperatures. You will classify some spectra, using a comparison set of spectra. You will also note the peak wavelengths of the spectra you classify.

INTRODUCTION

Understanding what information can be gained from observing the light emitted from objects in space is vital to astronomy. One very important phenomenon of light that will be used in this exercise is the spectrum: light from an object that has been spread out by wavelength, similar to the way light can be spread out by wavelength when it passes through a prism or is refracted and reflected through raindrops. There are two very distinct properties of a spectrum that will be important in this exercise: energy distribution and absorption lines.

The energy distribution curve of a star is sometimes called a blackbody curve because the shape of its energy distribution curve is much like that of a theoretical blackbody. A theoretical blackbody is a substance that absorbs all light incident upon it and re-emits that light at all wavelengths, with the peak wavelength of that emission corresponding to the temperature of the blackbody. You will be examining blackbody curves for several stars to determine the peak wavelength of that curve and, thus, the temperature of the star that emitted that light.

Absorption lines are dark lines for which much less light is observed. These lines occur at discrete wavelengths within the spectrum of a star. Much less light is present at the wavelengths where these lines occur because that light was absorbed by an intervening gas. In the case of a star, the intervening gas is the atmosphere of the star. So, the absorption lines in the spectrum of a star's light are due to materials found in the atmosphere of the star.

1. **Download and save or print out comparison spectra.** The Astrophysics Data System (ADS) is a resource that houses electronically scanned refereed publications in astronomy (http://adsabs.harvard.edu/abstract_service.html). You may find the article written by Jacoby, Hunter, and Christian in 1984, at the following url: http://adsabs.harvard.edu/abs/1984ApJS...56..257J. From here, click on the **Send PDF** button, where you can download the article. The spectra you are interested in begin on page 259.

2. **Describe the differences of the main spectral types.** Considering only the letters that denote the spectral class (O, B, A, F, G, K, M), determine the characteristics that separate one class from another. Be sure to record both peak wavelength and the wavelengths of several (at least three) of the most prominent absorption lines. (If the spectrum has many absorption lines very close together, you may record a band instead of one of the prominent absorption lines.) Use the Notes section found at the end of this exercise to record your observations.
3. **Determine the approximate surface temperature of the main spectral types.** Estimate the peak wavelength for each spectral type, and use Wein's law $\left(T = \frac{2.898 \times 10^{-3}}{\lambda_{max}}\right)$ to calculate the blackbody temperature that corresponds to that peak wavelength.
4. **Go to the ELODIE Web site to view stellar spectra. (http://atlas.obs-hp.fr/elodie/)** Look up and view the spectra of at least three stars from the list below. Identify the spectral type of each star by comparing the spectrum on ELODIE with the library from the article found in step 1. Be sure to note both the peak wavelength of the curve and several (at least three) absorption lines you notice. Determine which spectral classification is appropriate by matching the spectrum, and calculate the blackbody temperature by estimating the peak wavelength and using Wein's law.

HD191984	HD209747	HD212754
HD194244	HD210418	HD212943
HD206778	HD211976	

5. **Use WorldWide Telescope software to find each star whose spectrum you identified.** Use the research button to find out the correct spectral type for the star by checking the spectral type on SIMBAD.

OPTIONAL Step 6. Write a paper about the determination of spectral classes of stars. Do some research on the spectra of stars and how the spectra are related to the chemical composition of stars. Start with your textbook and use any other resources your instructor suggests or provides. Write about what you learn from this research. Include anything you learn about the history of stellar spectral classification.

NAME ____________________

DATE ____________________

INSTRUCTOR ____________________

SECTION ____________________

NOTES FOR What Are Stars Made Of?

1. Use this space to describe, in detail, the spectra for each spectral class (O, B, A, F, G, K, M). Be sure to note at least three major absorption lines, and try to estimate the peak wavelength, even if you can't see that part of the spectrum.

   O

   B

   A

   F

   G

   K

   M

2. Use this space to calculate approximate surface temperatures for each of the spectral class you noted above. Use Wein's law to calculate the temperature from your estimated peak wavelength.

O

B

A

F

G

K

M

3. Use this space to record your observations of the stars listed.

Name	Peak Wavelength	Temperature (Calculated Using Wein's law)	At Least Three Prominent Absorption Line Wavelengths	Your Spectral Classification
HD191984				
HD194244				
HD206778				
HD209747				
HD210418				
HD211976				
HD212754				
HD212943				

4. Use this space to record the accepted spectral classification that you find on SIMBAD for each star.

Name	Spectral Classification
HD191984	
HD194244	
HD206778	
HD209747	
HD210418	
HD211976	
HD212754	
HD212943	

5. Use this space to discuss any differences between your values and the values you found on the Web site. If your classifications are different from the accepted values, can you see why?

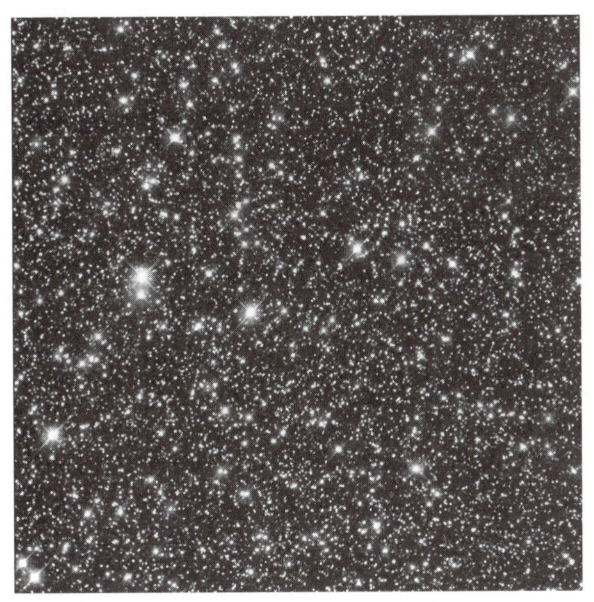

The Sagittarius star field. In this image, main sequence stars, red dwarfs, red giants, blue giants, white dwarfs, and brown dwarfs are visible.

9
Inferring Physical Properties

DESCRIPTION

Because stars are so distant, astronomers cannot directly measure physical properties of stars such as size, temperature, or mass. To learn about the physical nature of stars, it is necessary to infer this information from observations. To do this, astronomers rely on what is already known about objects that can be studied in the laboratory.

Using mathematical relations (laws) that have been derived from theory or laboratory experiments, astronomers can infer some of the physical properties of the objects they study. Measurements of brightness and calculations of distance can be made. These measurements can be combined to derive luminosity (the amount of energy emitted by a star). A star's spectrum can be analyzed to determine its spectral type, which is related to its surface temperature, through our understanding of how atoms behave at different temperatures.

This exercise uses real data (luminosity and spectral type) of stars and employs the Stefan-Boltzmann law for blackbodies to infer size information about the nearest and brightest stars.

INTRODUCTION

In an attempt to understand the physical nature of stars, astronomers have devised clever methods. One such method involves the use of a mathematical relationship that describes the behavior of blackbodies and, arguably, the most important analytical tool in astronomy, the H-R diagram.

The H-R diagram is a graph of the stars that uses spectral type on the horizontal axis and the luminosity on the vertical axis. It seems simple enough, but when this is done for, say, the 50 brightest stars in the sky and the 50 nearest stars, much more information about the physical nature of stars is revealed.

1. **Use the data from Table 1 at the end of this exercise to create a graph of the 50 nearest stars using Excel.** Enter the data into a worksheet on Excel, and then plot the data points into a scatter graph using the graph application. Use temperature for the *x*-axis, and make sure to force it to run backward so that the smallest numbers are on the left and the largest numbers are on the right. Then plot the luminosity (in terms of the solar luminosity) on the vertical axis.
2. **Notice that this graph appears to contain two distinct groupings of stars on the diagram.** Using the Stefan-Boltzmann relationship ($L \propto R^2 T^4$), determine the relative sizes of the two groups you identified. Which group must contain larger stars? Explain your reasoning for this conclusion.

(It is *not* necessary to make calculations for this part. Try to reason it through with rough numbers, if you need numbers. Look at stars with the same temperatures but different luminosities and stars with similar luminosities but different temperatures.) Use the Notes section found at the end of this exercise to record your responses to these questions.

3. **Now use the data from Table 2 at the end of this exercise to create a graph of the 50 brightest stars using Excel.** Plot these stars as a separate data set on the same graph. You may have to adjust the *y*-axis range to include all the data from both groups.

4. **With the addition of the brightest stars, you should be able to notice another grouping of stars (different from before)**. Also, one of the two original groups should be extended on the graph. Again, using the Stefan-Boltzmann relationship, ($L \alpha R^2 T^4$), determine the relative sizes of the new group to the two groups you already identified. Which group must contain the largest stars? Explain your reasoning for this conclusion. (It is *not* necessary to make calculations for this part. Try to reason it through with rough numbers, if you need numbers. Look at stars with the same temperatures but different luminosities and stars with similar luminosities but different temperatures.)

OPTIONAL STEP 5. Write a paper describing the relationship you used to determine the relative sizes of the stars from the luminosity and spectral type.

OPTIONAL STEP 6. Read about the history of the H-R diagram and include this information in your final report.

Table 1 Fifty Nearest Stars

Name	M_V	L/L_{Sun}	Spectral Class	T(K)
Proxima Centauri	15.53	5.25×10^{-5}	M5.5 eV	2940
Alpha Centauri A	4.37	1.53×10^{0}	G2 V	5830
Alpha Centauri B	5.72	2.27×10^{0}	K0 V	5240
Barnard's Star	13.23	4.36×10^{-5}	M5 V	3030
Wolf 359	16.57	2.01×10^{-5}	M6.5 Ve	<2850
Lalande 21185	10.46	5.60×10^{-3}	M2 V	3530
Sirius A	1.45	2.25×10^{1}	A1 Vm	9145
Sirius B	11.34	2.49×10^{-3}	A2	8810
Luyten 726-8A	15.42	5.80×10^{-5}	M5.5 de	2940
Luyten 726-8B (UV Ceti)	15.38	6.02×10^{-5}	M6 Ve	2850
Ross 154	13.14	4.74×10^{-4}	M3.6 Ve	3300
Ross 248	14.77	1.06×10^{-4}	M5.5 Ve	2940
Epsilon Eridani	6.15	2.96×10^{-1}	K2 V	5010
Ross 128	13.48	3.47×10^{-4}	M4+ V	3180
Luyten 789-6	14.63	1.20×10^{-4}	M5-M7 Ve	2850
Epsilon Indi	7.00	1.36×10^{-1}	K4/5 V	4450
61 Cygni A	7.50	8.55×10^{-2}	K5 V	4340
61 Cygni B	8.33	3.98×10^{-2}	K7 Ve	4040
Procyon A	2.67	7.31×10^{0}	F5 IV-V	6530

(continued)

Table 1 *(continued)*

Name	**M_V**	**L/L_{Sun}**	**Spectral Class**	**T(K)**
Procyon B	13.00	5.393×10^{-4}	A9 VII	7740
G 227-046	13.00	5.39×10^{-4}	M3.5 d	3280
Groombridge 34	11.18	2.88×10^{-3}	M2 V	3530
Lacaille 9352	9.56	1.28×10^{-2}	M2 V	3530
TAU Ceti	5.71	4.45×10^{-1}	G8 V	5430
G 051-015	17.01	1.34×10^{-5}	M6.5 eV	<2850
Luyten 725-32 (YZ Ceti)	14.20	1.79×10^{-4}	M5.5 Ve	2940
G 089-019	11.94	1.43×10^{-3}	M4 V	3180
Lacaille 8760	8.74	2.73×10^{-2}	M0 Ve	3800
Kapteyn's Star	10.94	3.60×10^{-3}	M1 VIp	3680
Kruger 60 A	11.85	1.56×10^{-3}	M2 V	3530
Kruger 60 B	11.59	1.98×10^{-3}	M4 V	3180
Ross 614	13.08	5.01×10^{-4}	M4.5 Ve	3105
G 153-058	12.02	1.33×10^{-3}	M2 d	3530
Van Maanen	14.21	1.769×10^{-4}	01 VII	200000
G 012-043 A	14.23	1.74×10^{-4}	M5.5+ Ve	2940
G 012-043 B	14.83	1.00×10^{-4}	M5.5 Ve	2940
L 1159-16	14.02	2.11×10^{-4}	M4.5 Ve	3105
L 143-23	15.61	4.87×10^{-5}	M4	3180
LP 731-58	17.32	1.009×10^{-5}	M6.5 V	2775
G 208-044	15.12	7.65×10^{-5}	M6 Ve	2850
G 267-025	10.25	6.79×10^{-3}	M2 V	3530
G 240-063	10.88	3.80×10^{-3}	M3.5 V	3280
L 145-141	13.14	4.74×10^{-4}	G6	5620
CPD-46 8664	11.05	3.25×10^{-3}	M3 d	3380
CI 20-1290	10.33	6.31×10^{-3}	M2 V	3180
G 158-027	15.38	6.02×10^{-5}	M5-5 V	2940
G 156-057	11.80	1.63×10^{-3}	M5 d	3030
CI 20-1046	12.58	7.94×10^{-4}	M3.5 d	3280
Groombridge 1618	8.19	4.53×10^{-2}	K7 V	4040
G 054-023	10.95	3.56×10^{-3}	M3.5 eV	3280

Table 2 Fifty Brightest Stars

Name	M_V	L/L_{Sun}	**Spectral Class**	T**(K)**
Sirius	–1.46	3.28×10^2	A1 Vm	9145
Canopus	–0.72	1.66×10^2	F0 II	7020
Arcturus	–0.04	8.87×0^1	K1.5 III	5068
Rigil Kentaurus	–0.01	8.63×10^1	G2 V	5830
Vega	0.03	8.32×10^1	A0 V	9480
Capella	0.08	7.94×0^1	G5 III	5680
Rigel	0.12	7.66×10^1	B8 I	11800
Procyon	0.38	6.03×10^1	F5 IV-V	6530
Achernar	0.46	5.60×10^1	B3 V	19000
Betelgeuse	0.50	5.40×10^1	M1 I	3680
Hadar	0.61	4.88×10^1	B1 III	25600
Altair	0.77	4.21×10^1	A7 V	7930
Aldebaran	0.85	3.91×10^1	K5 III	4340
Antares	0.96	3.53×10^1	M1 Ib	3680
Spica	0.98	3.47×10^1	B1 III	25600
Pollux	1.14	2.99×10^1	K0 III	5240
Fomalhaut	1.16	2.94×10^1	A3 V	8593
Mimosa	1.25	2.70×10^1	B0.5 IV	27650
Deneb	1.25	2.70×10^1	A2 Ia	8810
Acrux	1.33	2.51×10^1	B0.5 IV	27650
Regulus	1.35	2.47×10^1	B7 V	13000
Adhara	1.50	2.15×10^1	B2 II	22300
Gacrux	1.63	1.91×10^1	M3.5 III	3280
Shaula	1.63	1.91×10^1	B2 IV	22300
Bellatrix	1.64	1.89×10^1	B2 III	22300
El Nath	1.65	1.87×10^1	B7 III	13000
Miaplacidus	1.68	1.82×10^1	A2 IV	8810
Alnilan	1.70	1.79×10^1	B0 Ia	29700
Al Na'ir	1.74	1.72×10^1	B7 IV	13000
Alioth	1.77	1.68×10^1	A0	9480
Gamma2 Vela	1.78	1.66×10^1	O9 I	33200
Mirfak	1.79	1.64×10^1	F5 Ib	6530
Dubhe	1.79	1.64×10^1	K0 IIIa	5240
Wezen	1.84	1.57×10^1	F8 Ia	6137
Kaus Australis	1.85	1.56×10^1	B9.5 III	10090
Avior	1.86	1.54×10^1	K3 III	4785

(*continued*)

Table 2 (*continued*)

Name	M_V	L/L_{Sun}	Spectral Class	T(K)
Alkaid	1.86	1.54×10^1	B3 V	19000
Sargas	1.87	1.53×10^1	F1 II	6885
Menkaliman	1.90	1.49×10^1	A2 IV	8810
Atria	1.92	1.46×10^1	K2 IIb	5010
Alhena	1.93	1.45×10^1	A0 IV	9480
Peacock	1.94	1.43×10^1	B2 IV	22300
Delta Vela	1.96	1.41×10^1	A1 V	9145
Mirzam	1.98	1.38×10^1	B1 II	25600
Castor	1.98	1.38×10^1	A1 V	9145
Alphard	1.98	1.38×10^1	K3 II	4785
Hamal	2.00	1.36×10^1	K2 III	5010
Polaris	2.02	1.33×10^1	F7 Ib	6240
Nunki	2.02	1.33×10^1	B2.5 V	20650
Deneb Kaitos	2.04	1.31×10^1	K0 III	5240

NAME________________________________

DATE________________________________

INSTRUCTOR________________________________

SECTION________________________________

NOTES FOR Inferring Physical Properties

1. Attach your H-R diagram of the 50 nearest stars here.

2. Use this space to explain the relationship between size, temperature, and luminosity that you observe in your H-R diagram of the 50 nearest stars. Between the two groups represented on your graph, which contains the larger stars? Why?

3. Attach your H-R diagram including the 50 brightest stars here.

4. Use this space to explain the relationship between size, temperature, and luminosity that you observe in your H-R diagram of the 50 brightest and 50 nearest stars. Looking at the new grouping of stars, what must their sizes be relative to the other two groups? Explain how you know this.

The Pleiades cluster of stars.

10 Reading the Stars

DESCRIPTION

Stars come in many sizes and colors. It turns out that knowing a star's color and size can tell you where it is in its evolutionary cycle. Stars evolve from protostars to stellar remnants on time scales of hundreds of millions to hundreds of billions of years. Astronomy has not been around that long—humans haven't been around that long! How can astronomers know that stars change on those time scales?

The answer is that astronomers came up with a clever experiment to determine this information. This exercise presents you with the information necessary to deduce the basic outline of stellar evolution. Using astronomical data, you will analyze plots to determine which stages come first and how a star changes on the H-R diagram throughout its existence.

You will repeat a clever experiment. The purpose of repeating this experiment is manyfold. First, science requires that any experiment be repeatable; that is, no matter who does the experiment, the same result occurs. Second, repeating this experiment will deepen your understanding of the astronomy concepts and science processes involved in the experiment.

INTRODUCTION

In order to piece together the story of stellar evolution, you have to have all the data. In this exercise, the focus is on analysis of data, so data will be presented and explained to you, and you will have to analyze and interpret the data to draw conclusions about stellar evolution.

1. **Examine color-magnitude diagrams of clusters of stars.** Because a cluster of stars is a group of stars that were formed at the same time from the same cloud of gas and dust, we can learn a lot about the stars within that cluster. A color-magnitude diagram is a kind of H-R diagram. The horizontal axis is for the color, which is correlated to temperature, as on the H-R diagram. The vertical axis is for apparent magnitude, which is similar to absolute magnitude or luminosity, as on the H-R diagram. Because all the stars in a cluster are at the same location in space, the apparent magnitude (how bright a star appears) can be treated as a kind of absolute magnitude (how bright a star would appear if it were at a standard distance from Earth). At the end of this exercise, you will find modified color-magnitude diagrams (the vertical axis has been changed to absolute magnitude for comparison purposes) for several clusters of stars: the Pleiades, M11, Praesepe, and M67.
2. **Identify trends you see in the color-magnitude diagrams.** Do you see any trends among these color-magnitude diagrams? These diagrams have been modified so that the vertical axis is the absolute magnitude rather than apparent magnitude so that they can be compared. These four clusters were not

all formed at the same time. In fact, they have distinct ages that are quite different from one another. Can you determine the sequence from youngest to oldest? What do you notice about the main sequence, red giants, and white dwarfs on these diagrams. Can you put them in some order that makes sense based on what you see? Once you have put the diagrams together, compare your order to other groups in your class. If your order is different from that of another group, discuss to determine which way makes more sense. Try to come to consensus on an order and an explanation for your order. When you are confident that your ordering is complete, show your instructor. Use the Notes pages found at the end of this exercise to record your rationales and consensus sequence.

3. **Based on the order you have derived, what can you say about what happens to a star during its existence?** Does a star begin as a white dwarf, main sequence, or red giant star? Following that stage, what does the star become? Do all stars evolve the same way, at the same pace? How can you tell which stage comes first? When you have arrived at a sequence of stages for stellar evolution, run this by your instructor.
4. **Write up a summary of the sequence you have derived from the data.** Include any other information you know about these various stages in stellar evolution: the type of nuclear fusion going on in the core (if at all), the size, and the temperature, relative to the previous stage.

OPTIONAL STEP 5. Create a story of the life cycle of a star based on your findings. Be as creative as you want. Feel free to take artistic license (by making the stars anthropomorphic, for example), but include actual accurate scientific information (such as which stage comes first, relative temperatures and sizes, etc.).

COLOR MAGNITUDE DIAGRAMS

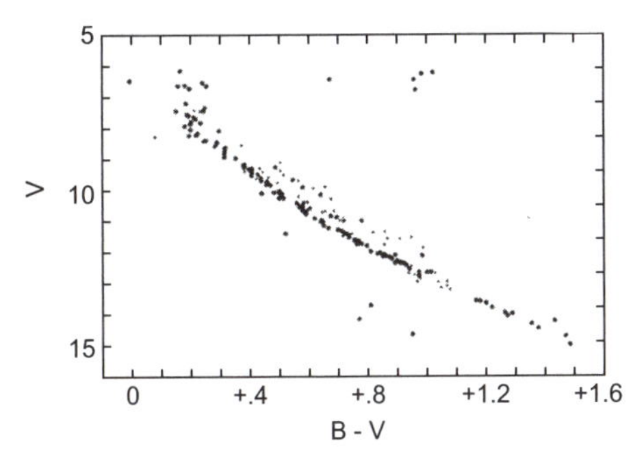

Fig 10.1 *h + x Persei*

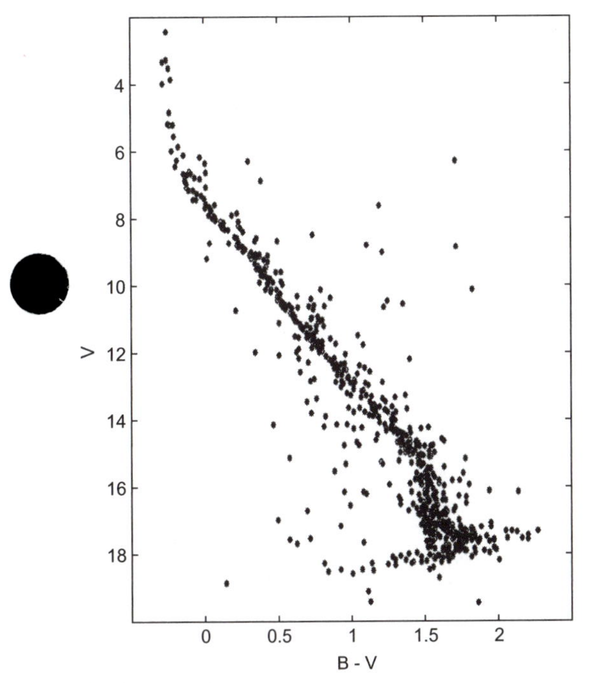

Fig 10.2 *Pleiades*

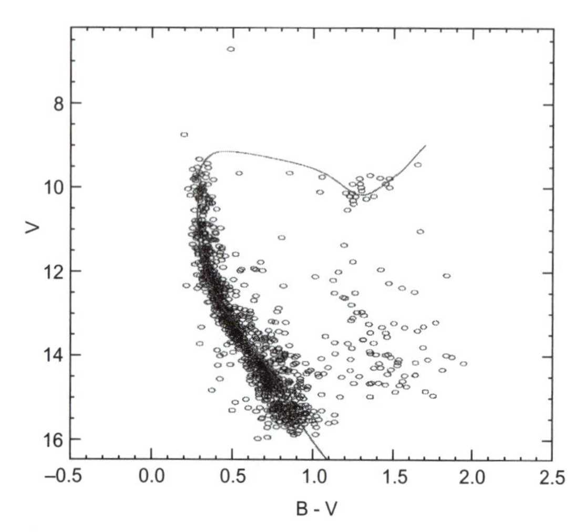

Fig 10.3 *Pleiades*

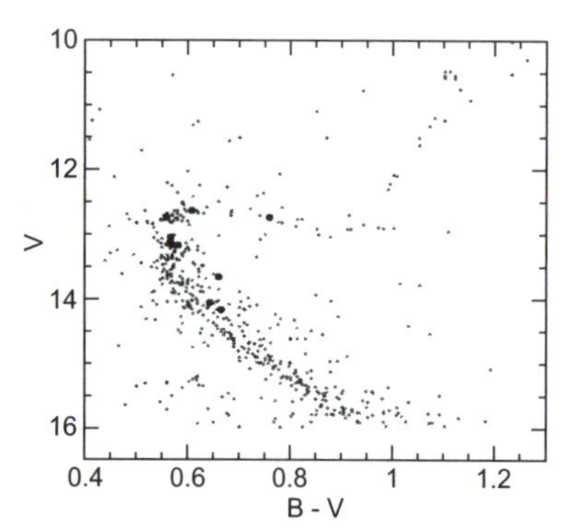

Fig 10.4 *M67*

NAME ______________________________

DATE ______________________________

INSTRUCTOR ______________________________

SECTION ______________________________

NOTES FOR Reading the Stars

1. Use this space to describe any trends you see in the color-magnitude diagrams.

2. Use this space to describe the order in which you placed the color-magnitude diagrams and explain why you chose this order.

3. Based on the order in which you put the color-magnitude diagrams, explain stellar evolution sequences. How does a star start out, and how does it change over time in luminosity and temperature?

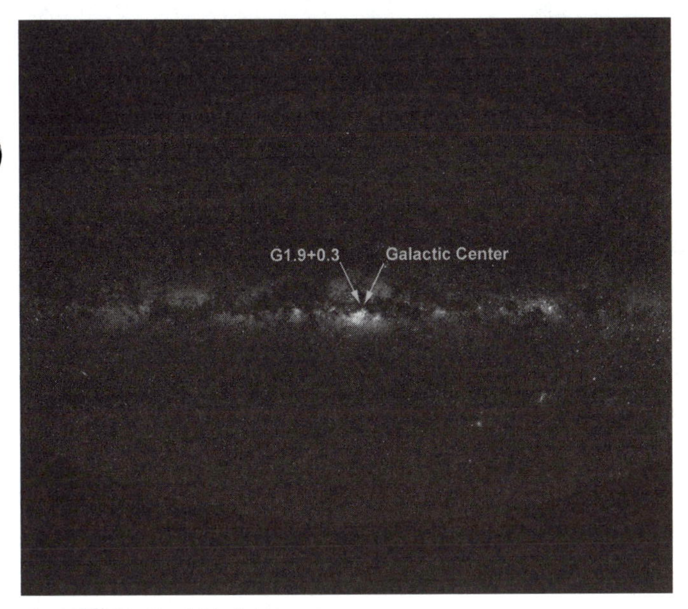

The Milky Way in visible light.

11
What Is the Milky Way Made Of?

DESCRIPTION

Galaxies, such as the Milky Way, are made of stars, gas, and dust. These constituents combine in different ways to make HII regions, planetary nebulae, reflection nebulae, supernova remnants, and dark nebulae. These objects have distinct, measurable properties that make them distinguishable. Each of these properties is related to the physical phenomenon that causes the object's appearance, although sometimes the name of the type of object is simply misleading. One important aspect of science is the ability to correctly classify objects. The first step toward being able to understand the physical nature of an object is to group it with other, similar objects so that one can learn as much as possible about the physical phenomenon that causes the appearance of the object.

In this exercise, you will use the WorldWide Telescope software to identify the type of object you are observing and list the object's intrinsic properties. Then, you will be asked to determine which physical phenomena are causing the appearance of each type of object, based on the information you find.

INTRODUCTION

Each of the objects you will observe fits into at least one of the following categories: HII region, planetary nebula, reflection nebula, supernova remnant, or dark nebula. You will be able to identify the type of object you are looking at by clicking on the Research button when you find the object on the WorldWide Telescope. You will also need to find out information about what color the object appears to be and what type of stars (if any) are associated with the object.

Each of the objects you will observe appears the way it does for one of the reasons listed below. Your task is to identify what type of object you are observing, describe the appearance of the object, and match up the type and appearance of each object with the description of the physical phenomenon that causes its appearance.

1. The blue light from the young, yellow and blue stars is being reflected from the dust behind them, making them look like they have a blue gas surrounding them.
2. Hydrogen gas, which was once part of a red giant, is cooling and expanding from a hot white dwarf core.
3. Light from behind this type of object is obscured from our point of view. When looking at these objects in the infrared, sometimes the stars behind the object are visible as faint, red light sources.

4. Hot gases, which used to be part of a massive star, are expanding at extremely high speeds out into space. The wispy, colorful gases are not associated with any stars, yet emit light and appear bright. At the center of this type of object, there may be a black hole or a pulsar.
5. Hydrogen gas surrounding the new, young, hot stars is excited, causing it to emit light.

List of objects:

- M45
- M43
- IC434
- M57
- M1

1. **Find each of the above objects using the WorldWide Telescope (WWT).** When you have found each object, use the finder scope feature to determine the classification of the object. This should match up with at least one of the five possible types of objects listed above (HII region, planetary nebula, reflection nebula, supernova remnant, or dark nebula). *You may have to move the finder scope around on the image to make sure you have the correct identification.* Record the information you gather in the Notes pages found at the end of this exercise.
2. **Describe the object you are observing.** Be sure to note the color of the object, the types of stars associated with the object, if applicable, and any shape to the object that you notice. If there are other features that seem unique to the object, write those down, too.
3. **Match your description to one of the physical phenomena in the Introduction.** You may need to use the research feature to find out information about the object, but you will probably be able to look at the image displayed on the WWT and be able to match the physical phenomenon with your description. Write down what you found out about the object that made you match it to the description you chose.

OPTIONAL Step 4. Find out more about this class of object. Use the Internet or your textbook to find out more about this type of object. Use the research feature on WWT to look up information on this object using one of the astronomical Web sites (NED, SIMBAD, or SEDS). You may find other images of the object or other information that you may determine to be relevant. Include references for all information you find.

OPTIONAL Step 5. Write a paper describing the procedure you used to classify objects and what you learned about how the names of some objects reveal their physical nature, whereas others are misleading. Include printouts of the images you used in your classification. Make sure to include references for these images as well as for any other information you used to classify these objects.

OPTIONAL EXTENSION. Find one more object for each category by searching with the WWT by category of object. Describe how the object appears, what color it is, its shape, and explain its appearance based on its classification.

NAME ______________________

DATE ______________________

INSTRUCTOR ______________________

SECTION ______________________

NOTES FOR What Is the Milky Way Made Of?

1. Fill in the table below:

Name	Type	Description	Physical Phenomenon
M45			
M43			
IC434			
M57			
M1			

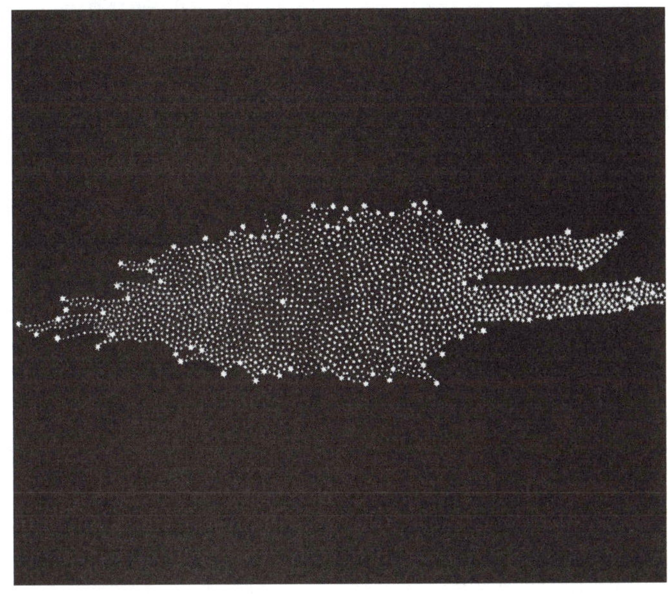

This image shows the result of the research William and Caroline Herschel did to try to determine the extent of the universe.

12 Where in the Milky Way Are We?

DESCRIPTION

In the 21st century, most of us are aware that Earth is one of several planets in our solar system, which is a system of planets that orbit a star we call the Sun. The Sun is one of several hundred billion stars in our galaxy, which we call the Milky Way; but where in the Milky Way are we?

At night, all the stars you see are part of the Milky Way. You may also know that the Milky Way is a spiral galaxy. How could we find out where Earth and the solar system are located within this spiral galaxy? This exercise gives students the tools to design and execute an experiment to determine our location within the Milky Way.

INTRODUCTION

Figuring out where we are in the universe is an important part of understanding why we observe what we do. From the beginning of recorded history, humans have tried to determine their place in the universe. The ancient Mayans thought that the universe consisted of Earth, the Sun, the Moon, the planets (Venus and Mercury), and the stars. Of course, Earth was at the center of their conception of the universe because it did not appear to move.

In the beginning of the 20th century, astronomers began to realize that the universe was much bigger than previously thought. This is when astronomers began to understand that we live in an enormous galaxy that is just one of billions of other galaxies in the universe. Until this realization, models of the universe did not go very far beyond the solar system. Most models assumed that outside the solar system there were stars. The rest of the universe was just more stars.

Now we know that stars exist primarily within structures known as galaxies. Galaxies contain stars (from tens of millions in dwarf galaxies to several trillions in the largest galaxies), gas, and dust. Our Sun is just one of several hundred billion stars in the Milky Way Galaxy. But, like a tree in a forest, it is hard to tell where we are. Given our history as humans, we would probably be satisfied with the premise that the Sun must be at the center of the Milky Way, but there is observational evidence to the contrary.

In this exercise, you will determine the (three-dimensional) shape of the Galaxy and our position within it (or at least, relative to its center).

PART I: THE SHAPE OF THE MILKY WAY

1. **Look up the morphological classification of the Milky Way.** There is a bit more to it than "spiral." Look for information on morphological classification in textbooks and/or online. Find out the shape

of the Milky Way Galaxy. Use the Notes section found at the end of this exercise to record your findings and list the sources you used.

2. **Identify the parts of the Milky Way.** Some possible parts of a galaxy are: bulge, disk, halo, bar, or arm. Knowing the type of galaxy the Milky Way is, you should be able to identify the features the Milky Way has. Create a physical model of the Milky Way and identify its features. (A globe is an example of a physical model of Earth.)

3. **Use your planetarium software to observe the Milky Way from the Northern and Southern Hemispheres.** Based on your observations (without the aid of telescopes or other wavelengths of light outside the visible range), explain how these observations support the conclusion that we live in a galaxy shaped like your model. Furthermore, determine in what part of the galaxy the Sun must be located to see the Milky Way as we do.

PART II: MAPPING THE MILKY WAY WITH GLOBULAR CLUSTERS

Galaxies such as the Milky Way have objects called *globular clusters* that exist in the outer regions of the galaxy (or halo). Globular clusters are dense clusters of stars that are generally free of gas and dust. In galaxies such as the Milky Way, the gas and dust are located primarily in the disk. If we want to look for objects that we can see clearly (with no interference from gas and dust), we must observe objects in the halo. In the late 19th century, an astronomer named Harlow Shapley devised an experiment to determine the position of Earth in the universe using globular clusters. He chose these objects because of their apparent lack of interaction with the gas and dust in the Milky Way.

OPTIONAL Step 4. Explain how globular clusters could be used to determine the position of Earth within the Milky Way. Given what you already know about the part of the galaxy in which the Sun is located and the information about globular clusters, what information would you need to figure out the position of Earth? For about how many globular clusters do you think he would need to gather this information in order to determine the position of Earth? Describe the experiment and list the steps one would need to take.

OPTIONAL Step 5. As a group, devise a method for determining the position of Earth within the Milky Way using globular clusters. Decide how many globular clusters you want to observe. Decide what information you need in order to do the analysis. There may be many different ways of analyzing the data. There does not need to be group consensus for the analysis, but each method that is used should be discussed with the group.

ALTERNATIVE to Steps 4 and 5. Your instructor will lay out for you a method for determining the position of Earth within the Milky Way using globular clusters.

6. **Gather the data you need from sources on the Internet. You may use a planetarium software to aid in the identification of globular clusters.** Share the responsibility for gathering data with every classmate. (No one should be exempt from gathering data.) Each individual class member should analyze the data set on his or her own, but gathering the data can be done much more quickly if everyone gets the data for a number of globular clusters and then shares the data. In fact, this is often how data are gathered in astronomy, because no one individual can get all the data they need all the time, and certainly not all at once.

7. **Analyze the data.** Use the entire data set gathered by the class and do the analysis you choose. Determine the position of Earth within the Milky Way. You should be able to determine distance from the center of the galaxy and angular position from some reference point (a measure similar to longitude on the surface of Earth that is relative to the prime meridian).

OPTIONAL Step 8. Create a scientific poster presentation. Your final product will be a poster presentation. You will need to outline the procedures you used to gather information (and data), display your physical model, show your data, describe your analysis process, and present your result. Compare your results with known values (use a textbook or some other reliable resource to check your result). Explain any differences between your result and the accepted values in terms of the data you used, the analysis you used, and the assumptions you made.

NAME______________________________

DATE______________________________

INSTRUCTOR______________________________

SECTION______________________________

NOTES FOR Where in the Milky Way Are We?

1. Write down the morphological classification for the Milky Way and the source where you found this information.

2. List the parts of the Milky Way that you identified in your model.

3. Use this space to explain how your observations with the planetarium software support the conclusion that we live in a galaxy that has the shape of your model. Include where you think the Sun should be in your model to see the Milky Way as we do.

4. Describe the method you will use to determine the position of Earth within the Milky Way using globular clusters.

5. Use this space to tabulate the data you gather. You may print out and attach a table or graph, if you like.

6. Use this space to attach a graph of the data compiled by the entire class to determine the location of Earth in the Milky Way. Indicate Earth and the center of the Milky Way on the graphic. Include any calculations you made to determine the distance to the center of the Milky Way.

A cluster of stars in the nearby irregular galaxy, the Small Magellanic Cloud.

13
How Close Is Our Nearest Neighbor?

DESCRIPTION

From Shapley's experiment ("Where in the Milky Way Are We?" Exercise 12 in this book), we have learned that our Sun is located about halfway between the center of the Milky Way and its outer edge. Our Milky Way contains hundreds of billions of stars and is one of billions of galaxies in the universe. The universe is vast, and most of the universe is empty: no stars, no dust, no gas, no galaxies. Yet, our Milky Way is located in a group of about 35 galaxies. The galaxies in our group are so far apart that our closest neighbor is still hundreds of thousands of light-years away.

The distances between galaxies are so large that they cannot be measured using the same method for measuring the distances to stars within our galaxy. Paradoxically, the galaxies in a group are so close together that the expansion of the universe does not dominate their motion, so the Hubble law (used in "Our Expanding Universe," Exercise 15 in this book) cannot be used to measure the distances to galaxies that are members of the Local Group. To determine the distance to our nearest neighboring galaxies requires a clever astronomy experiment.

When Shapley did his experiment, he had to measure the distances to globular clusters. To do this, he used Henrietta Leavitt's discovery that certain variable stars, Cepheids, obeyed a period-luminosity law that showed that their luminosities could be determined by measuring their periods of variation. These variable stars have been found in other galaxies, including our nearest neighbors.

INTRODUCTION

Cepheid variable stars are simply stars whose brightness varies regularly. They are called Cepheids because the first such star was found in the constellation Cepheus. Henrietta Leavitt's work studying these stars showed that their periods of variation (how long it took them to go from maximum brightness to minimum brightness and back to maximum brightness again) were directly correlated to their luminosities. That is, the more luminous the star, the longer it would take for that star to change its brightness.

What is most important about this discovery is that it makes it possible to calculate the luminosity of a star without having to measure its distance. Instead, one can measure its period, find its luminosity, and then determine its distance from Earth, using the distance modulus formula. So, two Cepheids with the same period of variation will have the same luminosity, but if one appears fainter than another, the fainter star must be farther away.

In this exercise, you will derive the period-luminosity relationship, using modern data. Then, you will apply this law to a Cepheid variable star in the Large Magellanic Cloud, one of our nearest neighbors. Given its period, you can find its absolute magnitude with the period-luminosity law. Then, given its apparent magnitude, you can find its distance. Knowing the distance to this Cepheid will tell us the distance to the Large Magellanic Cloud.

1. **Plot the data from the table below, using Excel.** Plot log *P* on the *x*-axis and the absolute magnitude on the *y*-axis.
2. **Fit a line to the data.** Use the fit line (trend line) function to fit a line to the data and display the equation of the best fit line. The equation you see will tell you how *M*, the absolute magnitude, is related to log *P*. Use the Notes section at the end of this exercise to keep a record of your analysis.
3. **Calculate the distance to the Large Magellanic Cloud.** Use a Cepheid star with $P = 4.76$ days and apparent magnitude $(m) = 15.56$. Calculate log *P*. Restate the equation of your best fit line from step 2. The *y* is absolute magnitude, and the *x* is log *P*, the logarithm of the period. Use your value of log *P* and the best fit equation to calculate the absolute magnitude, *M*. Use your calculated value for absolute magnitude and the given apparent magnitude in the distance modulus formula to calculate the distance to the star in units of parsecs. The distance to the star is the same as the distance to the Large Magellanic Cloud (LMC).

   Distance modulus formula: $d = 10^{\left(\frac{m - M + 5}{5}\right)}$

4. **Compare the distance you calculated using the period-luminosity relationship that you derived to the accepted distance to the LMC.** Look up the distance to the LMC on the NASA/IPAC Extragalactic Database (NED), and compare the answer you got to the distance listed. Calculate the percent error. (For help on how to calculate this, see the appendix.)

OPTIONAL STEP 5. Write a paper describing how you derived a period-luminosity law and calculated the distance to the nearest galaxy. Include the comparison you made to the accepted distance to the LMC. Discuss any source of error that you can determine.

Data Table

Star Name	M	Period (days)	Log P (days)
RY Cas	−8.54	12.14	1.084
FM Cas	−5.83	5.81	0.764
VX Pup	0.53	3.01	0.479
FN Vel	−0.37	5.32	0.726
SZ Tau	−1.03	3.15	0.498
Alpha UMi	−3.60	3.97	0.599
delta Cep	−3.32	5.37	0.730
AP Vel	−2.49	3.13	0.496
U TrA	−1.52	2.57	0.410
UZ Cen	1.66	3.33	0.522

NAME________________________________

DATE________________________________

INSTRUCTOR________________________________

SECTION________________________________

NOTES FOR How Close Is Our Nearest Neighbor?

1. Attach your graph of the data here (after you have included your trend line and displayed the equation).

2. Use this space to calculate the distance to the Large Magellanic Cloud.

   a. First calculate log *P*.

   b. Calculate *M*, using your trend line equation and your calculation from 2.a above.

   c. Calculate the distance to the Large Magellanic Cloud using the apparent magnitude given and the absolute magnitude you calculated in 2.b above.

3. Find the accepted distance to the Large Magellanic Cloud and compare it to your distance. What is the percent error?

A cluster of galaxies known as Abell Cluster S0740. This cluster includes galaxies of many different morphologies.

14
The Galaxy Zoo

DESCRIPTION

Astronomy has not yet revealed much about the nature of galaxies. For the most part, astronomers know that galaxies are composed of stars, gas, and dust. They appear to have different star formation histories and differing amounts of stars, gas, and dust, but little is known about how galaxies get to be the way they are.

In order to better understand galaxies, however, astronomers have attempted to divide them into groups by their appearance. It is possible that the appearance of galaxies may be related to their history or evolutionary track. Although, as yet, there is not enough evidence to support this hypothesis, the first step in any endeavor to understand something scientifically is to classify and compare.

In this exercise, you will look at a number of galaxies and divide them into categories that you determine. Following this, you will compare how your classification scheme compares with the one astronomers have designed.

INTRODUCTION

Galaxies are large groups of gravitationally bound stars, with gas and dust clouds intermingled. In this exercise, you will use the WorldWide Telescope software to observe several galaxies. You will then divide these galaxies into at least two groups based on their shape and appearance. You will determine the nature of the categories that you develop. At the end of the exercise, you will compare the groupings that you made to those made by astronomers and discuss whether grouping galaxies in either way is helpful to understanding the nature of galaxies.

1. **Use the WorldWide Telescope to observe the galaxies listed in the table below.** Make sure that you look at each galaxy carefully. You may decide how to categorize the galaxies at any point during these observations, but you must designate a category for every galaxy in the table. Your categories may be broad or narrow. You must have at least two, but you should not have more than ten.
2. **Write a description of how you classified the galaxies you observed.** What features did you consider in classifying these galaxies? What features did you ignore? Did you develop a symbol or designation for your classifications? What do your symbols or designations mean? Use the Notes space found at the end of this exercise to record your responses to these questions.

List of Galaxies			
NGC 628	NGC 300	NGC 393	NGC 7331
NGC 4486	NGC 3031	NGC 205	NGC 523
NGC 1300	NGC 5457	NGC 4472	NGC 1566
IC 1623B	NGC 1316	NGC 4594	NGC 536
NGC 5194	NGC 620	ESO 286-19	

3. **Now go back and use the Research button for each galaxy to determine the astronomical classification of each.** Click the Research button in the pull-down menu and choose Information. The galaxy morphology classification will be most easily found using NED, the NASA/IPAC Extragalactic Database. The designations should be *S* for spiral, *SB* for barred spiral, *E* for elliptical, *I* for irregular, and *pec* for peculiar. If you are unsure how to interpret the classification for each galaxy, ask for help from your instructor.
4. **Now compare your classification scheme with the astronomical one.** Does either lend any insight into the nature of a galaxy? Is one better than the other?

OPTIONAL STEP 5. Look up other characteristics of these galaxies (such as color, size, number of satellites, number of arms, etc.) and see whether any of these are correlated to the morphological classifications (i.e., whether all spirals are large).

OPTIONAL STEP 6. Read a textbook or other reliable information source to learn more about the morphological classification of galaxies. Write a paper about the insight that has been gained from these designations, if any.

NAME ____________________

DATE ____________________

INSTRUCTOR ____________________

SECTION ____________________

NOTES FOR The Galaxy Zoo

1. Use this space to describe the categories you will use to classify the galaxies. You should have at least two categories, but no more than ten.

2. Use this space to classify the galaxies on the list.

   NGC 628

   NGC 300

   NGC 393

NGC 7331

NGC 4486

NGC 3031

NGC 205

NGC 523

NGC 1300

NGC 5457

NGC 4472

NGC 1566

IC 1623B

NGC 1316

NGC 4594

NGC 536

NGC 5194

NGC 620

ESO 286-19

3. Use this space to note the official classifications for the galaxies on the list.

NGC 628

NGC 300

NGC 393

NGC 7331

NGC 4486

NGC 3031

NGC 205

NGC 523

NGC 1300

NGC 5457

NGC 4472

NGC 1566

IC 1623B

NGC 1316

NGC 4594

NGC 536

NGC 5194

NGC 620

ESO 286-19

4. Use this space to discuss the usefulness of the two classification systems in learning about the physical nature of galaxies. Due to either classification of galaxies, do you know anything new about the physical nature of galaxies?

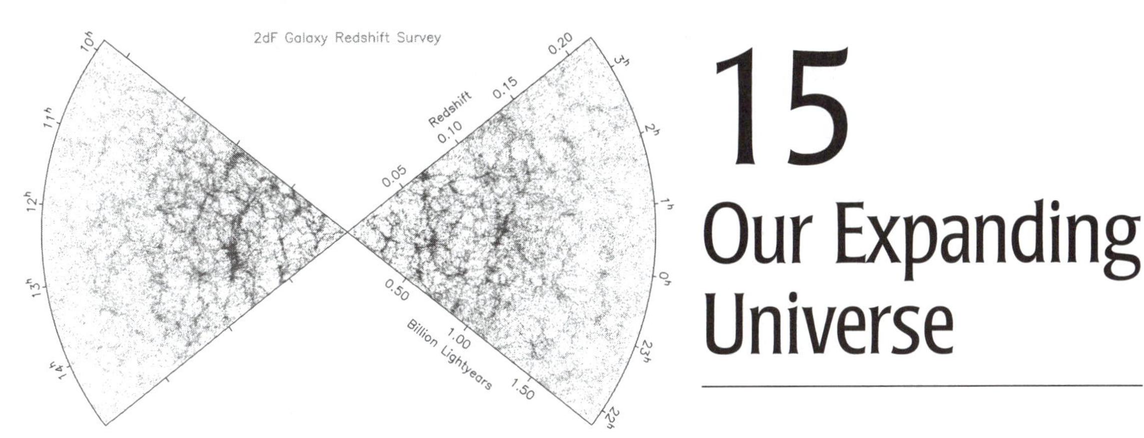

The mapping of the positions of thousands of galaxies. This redshift survey shows how matter in the universe is distributed.

15 Our Expanding Universe

DESCRIPTION

In the early 20th century, astronomers made an amazing discovery that changed everything they thought they understood about the universe. The discovery made was that almost every galaxy is moving away from us. Before this discovery, astronomers knew that the universe was full of stars and nebulae, but they had no idea that some of the nebulae were actually other galaxies or that these galaxies were moving away from us.

When Hubble and Humason first made this discovery, they plotted the distance to the galaxies on one axis versus the recessional speed of the galaxies on the other axis and found that they were correlated. The tight correlation implied a fundamental relationship, which led to the calculation of what is now known as the Hubble constant. The relationship is known as Hubble's Law. As any law in science, it merely describes the way the data behave and does not explain why. The explanation for why galaxies appear to be moving away from us at faster rates the farther they are away from us comes from the Big Bang theory. Theories seek to explain why, and laws simply describe repeated observable behavior.

According to the Big Bang theory, the Hubble constant describes the rate at which the universe is expanding. Hubble first determined the value of this constant to be 500 km/s/Mpc. This value means that for each megaparsec of distance from us an object is, its distance from us is increasing by 500 km every second (or, conversely, a megaparsec (3.086×10^{19} km) of space is increasing in size by 500 km every second). This exercise will give you a chance to understand how these measurements are made. In addition, you will calculate the value of the Hubble constant based on real data.

INTRODUCTION

Determining the rate of expansion of the universe is an endeavor that astronomers have undertaken for almost a century. For decades, there were two conflicting values for this number. Only recently have astronomers begun to come to agreement about the value of the Hubble constant, but it is not through the measurements of recessional speeds of galaxies. As will be evident with this experiment, there is good reason to be skeptical of the determination of Hubble's constant through a correlation of recessional speeds with distance.

To measure the recessional speed of a galaxy, first, you will acquire a galaxy spectrum, identify an absorption line, and measure the redshift of the galaxy. Next, you will use a simple equation to determine the recessional speed of the galaxy, and then you will compare your answer with the accepted value for recessional speed for that galaxy.

Finally, you will use NED (the NASA/IPAC Extragalactic Database) to acquire recessional speeds and distances for several (at least ten) galaxies of your choice. Using these data, you will determine the value for the Hubble constant and compare this value with Hubble's original value (500 km/s/Mpc) and the current accepted value (72 km/s/Mpc).

1. **Get the spectrum of a galaxy.** Go to http://nedwww.ipac.caltech.edu/ and select **Spectra** under the heading **Data**. Once there, enter **NGC1050** for the object name and click on the **Submit Query** button. The second spectrum that appears shows data for the wavelengths from 5445 to 7899 angstroms. (This information appears in the far right column of the table.)
2. **Determine the wavelength of the tallest peak in the spectrum.** Launch the Specview Applet by clicking on the **Specview** below the spectrum for NGC 1050. This applet will allow you to display the data using any units, to display reference lines, and to measure the wavelength of the emission lines shown. You may have to change the units displayed.
3. **Calculate the redshift of the galaxy.** Usually, the laboratory wavelength of the tallest peak in the spectrum is 6562.8 angstroms. The redshift of the galaxy is equal to the observed wavelength (what you measure) minus the laboratory wavelength, divided by the laboratory wavelength. The number you get should be a number between zero and 1, closer to zero. Round off your answer to the nearest one-thousandth. Use the Notes space found at the end of this exercise to do your calculations.
4. **Calculate the recessional speed of the galaxy.** The recessional speed of the galaxy is the redshift times the speed of light. Use the speed of light in kilometers per second so that your answer will be in those units. Check the answer you get against the value on NED. To check the value on NED, go to the Web site above and click on **Redshifts** under the heading of **Data**. Once there, enter **NGC 1050** for the object name and click on the **Redshift** button. Scroll down to find the recessional speed, which will be labeled **Velocity** or **Helio. Radial Velocity**. Ask your instructor which value you should use if the values are very different from one another.
5. **Look up the luminosity distance for the galaxy.** Go back to the Web site and look up the galaxy using the **By Name** option under the **Objects** heading by entering the galaxy name and clicking the **Submit Query** button. At the very bottom of the page, there is a section called **Cosmology-Corrected Quantities**. There should be at least three different distances listed. Use the Luminosity Distance. This distance is calculated using the luminosity and apparent brightness of the galaxy. This distance does not depend on the motion of the galaxy, so it is what we will use as an independent measure of distance.
6. **Determine the value of the Hubble constant.** Repeat step 5 and get the heliocentric radial velocities and luminosity distances for at least nine other galaxies from the list found at the end of this exercise. Make sure the velocities are all greater than 1000 km/s. (This is because the motions of galaxies that are moving at slower speeds are not dominated by the expansion of the universe.) To determine the value of the Hubble constant, plot the data you collected on a graph using Excel. Put the speed on the *y*-axis and the distance on the *x*-axis. Make sure speed is in units of kilometers per second and distance is in megaparsecs. The slope of the line that fits these data points is the Hubble constant.
7. **Compare the value of the Hubble constant you determined to the current accepted value (72 km/s/Mpc) and to Hubble's original value (500 km/s/Mpc).** Determine the percent difference between your value and the other two values. (The formula for percent difference can be found in the appendix.) The Hubble constant can also be used to calculate the age of the universe, assuming a constant rate of expansion. To perform this calculation, all you have to do is divide the number of kilometers in a megaparsec by the Hubble constant value, then divide it by the number of seconds in a year. That will give you the age of the universe in years. Use your textbook or the Internet to find these conversion values.

OPTIONAL STEP 8. For the final product of this exercise, write a report in the style of a scientific publication. You should have an abstract explaining the goal of the experiment and a brief description of the experiment and the results. The introduction should describe the history behind the experiment and methodology. Use your textbook and reliable Internet resources (such as NASA Web sites) as resources for this part. Be sure to address why one should be skeptical of the determination of Hubble's constant through a correlation of recessional speeds with distance. The next section of your report is the data and observations section in which you describe how you acquired the data for this experiment and what methods you used to acquire it. Here you should explain what types of spectra you used to get your recessional speeds; also, you should describe the process you used to extract the data for NGC 1050. The next section is the analysis section. Here you present the graph you made of the data you collected and describe the analysis you did to extract the Hubble constant. Your comparison to the currently accepted and original values should be included in this section, as well as your calculations of the age of the universe. Finally, the conclusions section will summarize the experiment and list the conclusions you drew from your data and your analysis. Don't forget to include a references section, and list all resources you used for this experiment.

List of Galaxies			
NGC47	NGC2139	NGC4023	NGC6023
NGC235	NGC2300	NGC4275	NGC6248
NGC498	NGC2514	NGC4498	NGC6545
NGC794	NGC2767	NGC4613	NGC6771
NGC926	NGC2906	NGC4832	NGC6917
NGC1065	NGC3089	NGC5006	NGC7015
NGC1238	NGC3333	NGC5176	NGC7222
NGC1567	NGC3554	NGC5443	NGC7525
NGC1723	NGC3725	NGC5624	NGC7765
NGC1956	NGC3952	NGC5863	NGC7823

NAME_______________

DATE_______________

INSTRUCTOR_______________

SECTION_______________

NOTES FOR Our Expanding Universe

1. Write down the wavelength of the tallest peak in the spectrum of NGC 1050.

2. Calculate the redshift of NGC 1050, given that the laboratory wavelength of the tallest peak in the spectrum is 6562.8 angstroms.

3. Calculate the recessional speed of NGC 1050 and compare it to the accepted value.

4. Write down the luminosity distance for NGC 1050.

5. Complete the table below:

Galaxy Name	**Heliocentric Radial Velocity (>1000 km/s)**	**Luminosity Distance (Mpc)**

6. Attach a copy of the Excel plot of the above data (adding in your data for NGC 1050).

7. Compare your value of the Hubble constant with the accepted value (72 km/s/Mpc) and to Hubble's original value of 500 km/s/Mpc. Calculate percent differences.

8. Calculate the age of the universe using your value of the Hubble constant.

Instructions for Using SkyGazer

The following list relates specific help for Exercises:

MOON PHASES AND SCIENTIFIC MODELS

Does not use SkyGazer.

MOTIONS IN THE NIGHT SKY AND THE CELESTIAL SPHERE

To set the sky to your location:

1. Click on the **Chart** pull-down menu.
2. Choose **Set Location**. Choose your state and city, and press **OK**.

To set to the time of your observation:

1. Click on the **Chart** pull-down menu.
2. Choose **Set Time**. Type in the date and/or time that you want.
3. Use the play icon above the clock in the **Time** window to see how the stars move.

THE SUN THROUGH THE SEASONS

To set the sky to January 1 of this year:

1. Click on the **Chart** pull-down menu.
2. Choose **Set Time**. Type in the date and/or time that you want.
3. Click the **OK** button.

To observe sunrise:

1. Follow the instructions above to set your location.
2. Click on the play icon to the right of the clock in the **Time** window and select **Sunrise**.
3. Use the mouse to move around until you are viewing the eastern horizon from your location.
4. Use the pause/play icon above the clock in the **Time** window to see the sky step forward in time. The Sun will be a bright object on the yellow ecliptic line (which can be turned on by clicking on the button that is third from the right-hand end of the **Display** window).

To observe rising constellation at sunset:

1. Clicking on a constellation will bring up an information window that will tell you the name of the constellation you clicked on.

MARS'S MOTION AND MODELS OF THE SOLAR SYSTEM

To observe the retrograde motion of Mars:

1. Find Mars by clicking on the **Center** pull-down menu, and click on **Find and Center**.
2. Type **Mars** into the text box and click on **Center**.

3. Click on the **Lock** button in the **Info** window that pops up.
4. Click on the play icon located above the clock in the **Time** window.

HOW DID GALILEO GO BLIND? OR, IT'S THE DATA, STUPID!

Does not use SkyGazer.

WHAT'S AN ASTRONOMICAL UNIT?

To view the transit of Venus:

1. Set the location to a city on a single line of longitude at a very high northern latitude.
2. Go to the position of the Sun using the **Center** pull-down menu, clicking on **Find and Center**, typing **Sun**, then clicking **Center**.
3. Use the + key to zoom in so you can see the entire disk of the Sun. (If you zoom in too far, use the – key to zoom back out.)
4. Set the date to June 8, 2004. and start the time at 12 hours, 0 minutes, 1 second A.M.
5. Set the time step to 10 seconds or less in the **Time** window and use the play icon located above the clock in the **Time** window to see Venus approach the disk of the Sun.

To change location:

1. Set the location to another city on the same line of longitude at a very low southern latitude.
2. Repeat the above steps to make the observations needed for the exercise.

HOW DO WE KNOW THE MASS OF JUPITER, ANYWAY?

To observe the motions of the Galilean satellites:

1. Go to the position of Jupiter using the **Center** pull-down menu, click on **Find and Center**, then type **Jupiter** and click on **Center**.
2. Zoom out and in using the + and – keys until you can see all four Galilean satellites.
3. To cause the satellites to move relative to Jupiter, use the play icon located above the clock in the **Time** window, with the time step set to anywhere from 1 to 24 hours.

To make measurements:

1. Pause the motion at the same time for each day or after a set number of hours and with a ruler measure the distance on the screen of each satellite from the apparent center of Jupiter.
2. Note the time and date (in the **Time** window) for each measurement you make.

WHAT ARE STARS MADE OF?

Does not use SkyGazer.

INFERRING PHYSICAL PROPERTIES

Does not use SkyGazer.

READING THE STARS

Does not use SkyGazer.

WHAT IS THE MILKY WAY MADE OF?

Does not use SkyGazer.

WHERE IN THE MILKY WAY ARE WE?

To view the Milky Way from the Northern and Southern Hemispheres:

1. Set your location in the Northern Hemisphere by clicking on the **Chart** menu and clicking on **Set Location**. Then, type the name of the city in the window, or pick a city from the alphabetical list or by clicking on a location on the globe shown.
2. Turn on the Milky Way by clicking on the sixth button from the left on the **Display** window. Zoom out (with the – key) and use the mouse to pan around until you see a large band of light—the Milky Way.
3. Set your location to a city in the Southern Hemisphere using the same method as in step 1.

SkyGazer cannot be used to complete the second part of this exercise.

HOW CLOSE IS OUR NEAREST NEIGHBOR?

Does not use SkyGazer.

THE GALAXY ZOO

Does not use SkyGazer.

OUR EXPANDING UNIVERSE

Does not use SkyGazer.

Instructions for Using Starry Night College

The following list relates specific help for Exercises:

MOON PHASES AND SCIENTIFIC MODELS

To observe the phases of the Moon within an accurate model:

1. On the top menu bar, select **Favorites**.
2. On the pull-down menu, select **Earth, Moon and Sun**, **Moon** and **Phases from Space**.
3. Press the **Play** button to observe the Moon moving around Earth and see the phases as viewed from Earth.

MOTIONS IN THE NIGHT SKY AND THE CELESTIAL SPHERE

To set the sky to your location:

1. Click on the **Location** window in the menu at the top of the screen and choose the **Other** option, if your location is not correct.

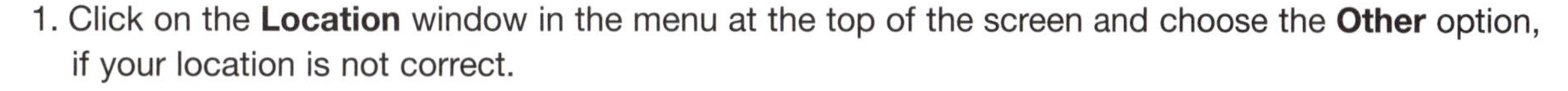

To set to the time of your observation:

1. Click on the hour and minute display shown in the upper left-hand corner, and type in the time you want.
2. Set the rate of time to 300 × and click on the **Play** button.

THE SUN THROUGH THE SEASONS

To set the sky to January 1 of this year:

1. Click on the month, day, and year on the top menu, and type in the month, day, and year you want.

To observe sunrise:

1. Follow the instructions above to set your location.
2. Click on the screen and scroll around along the horizon until you can see the **E** indicating the direction east.
3. Use the fast-forward and/or rewind buttons in the menu that pops up horizontally from the lower left-hand corner of the screen until you see the Sun rising near the **E**.

To observe rising constellation at sunset:

1. Turn on the constellation names and constellation outlines by clicking the first two icons that appear in the horizontal menu that pops up from the lower left-hand corner of the screen.

MARS'S MOTION AND MODELS OF THE SOLAR SYSTEM

To observe the retrograde motion of Mars:

1. On the top menu bar, select **Favorites**.
2. On the pull-down menu select **The Planets**, **Inner Planets**, **Mars Retrograde**.

HOW DID GALILEO GO BLIND? OR, IT'S THE DATA, STUPID!

Does not use Starry Night College.

WHAT'S AN ASTRONOMICAL UNIT?

To view the transit of Venus:

1. Set the location to a city on a single line of longitude at a very high northern latitude.
2. Go to the position of the Sun using the **Find** tab, click on **Sun**, and then click the play symbol and choose to magnify the Sun.
3. Use the **Options** tab to turn off daylight, if necessary.
4. Set the date to June 8, 2004, and start the time at 12:00:00 A.M.; use the fast-forward button to see Venus approach the disk of the Sun.

To change location:

1. Set the location to another city on the same line of longitude at a very low southern latitude.
2. Repeat the above steps to make the observations needed for the exercise.

HOW DO WE KNOW THE MASS OF JUPITER, ANYWAY?

To observe the motions of the Galilean satellites:

1. Find Jupiter by clicking on the **Find** tab and selecting **Jupiter**. When you press the play icon, choose **Magnify** to zoom in on Jupiter.
2. Zoom out and in using the + and – keys until you can see all four Galilean satellites. You can verify that the starlike objects are moons by moving the mouse over them. The name and type of object should appear as you move over each object.
3. Set the time steps to 1 day and use the pause/forward button to step forward in time steps of 1 day.

To make measurements:

1. Pause the motion at the same time for each day or after a set number of hours and with a ruler measure the distance on the screen of each satellite from the apparent center of Jupiter.
2. Note the time and date for each measurement you make.

WHAT ARE STARS MADE OF?

Does not use Starry Night College.

INFERRING PHYSICAL PROPERTIES

Does not use Starry Night College.

READING THE STARS

Does not use Starry Night College.

WHAT IS THE MILKY WAY MADE OF?

Does not use Starry Night College.

WHERE IN THE MILKY WAY ARE WE?

To view the Milky Way from the Northern and Southern Hemispheres:

1. Set your location in the Northern Hemisphere by clicking on the **View** pull-down menu and choosing your location.
2. Use the **Find** tab to locate the Milky Way by typing in *Milky Way* in the text box that appears.
3. Set your location to a city in the Southern Hemisphere using the same method as in step 1.

To find objects:

1. Click on the **Find** tab.
2. Type in the name of the object in the search window. (*Note:* Do not use spaces in the name you type).
3. Move the mouse over the object to get basic information about the object.

HOW CLOSE IS OUR NEAREST NEIGHBOR?

Does not use Starry Night College.

THE GALAXY ZOO

Does not use Starry Night College.

OUR EXPANDING UNIVERSE

Does not use Starry Night College.

Instructions for Using Stellarium

FOR MACS

1. Go to the Stellarium Web page (www.stellarium.org).
2. Download and install the Stellarium software for Macs.

FOR PCS

1. Go to the Stellarium Web page (www.stellarium.org).
2. Download and install the Stellarium software for PCs.

Following is specific help for exercises.

MOON PHASES AND SCIENTIFIC MODELS

Does not use Stellarium.

MOTIONS IN THE NIGHT SKY AND THE CELESTIAL SPHERE

To set the sky to your location:

1. Click on the **Location** window in the menu that pops up vertically from the lower left-hand corner of the screen.
2. In the top of the location window, next to the magnifying glass, type in the name of the city or town where you are and choose the correct option as it appears in the menu below that line.
3. You may select the box next to the words **Use a Default** so that you don't have to do this all the time.

To set to the time of your observation:

1. Click on the **Date/Time** window in the menu that pops up vertically from the lower left-hand corner of the screen.
2. Adjust the time in the window to match the time of your observation.
3. Use the fast-forward button in the menu that pops up horizontally from the lower left-hand corner of the screen.

THE SUN THROUGH THE SEASONS

To set the sky to January 1 of this year:

1. Click on the **Date/Time** window in the menu that pops up vertically from the lower left-hand corner of the screen.
2. Adjust the date in the window to January 1 of this year.

To observe sunrise:

1. Follow the instructions above to set your location.
2. Click on the screen and scroll around along the horizon until you can see the **E** indicating the direction east.

3. Use the fast-forward and/or rewind buttons in the menu that pops up horizontally from the lower left-hand corner of the screen until you see the Sun rising near the **E**.

To observe rising constellation at sunset:

1. Turn on the constellation names and constellation outlines by clicking the first two icons that appear in the horizontal menu that pops up from the lower left-hand corner of the screen.

MARS'S MOTION AND MODELS OF THE SOLAR SYSTEM

To observe the retrograde motion of Mars:

1. Find Mars by clicking on the magnifying glass tool in the menu that pops up vertically from the lower left-hand corner of the screen and typing **Mars** in the text window that pops up. Hit **Enter** and the software will center on Mars.

1. Use the **]** key or the **[** key to move forward or backward in time in steps of 1 week. Use **=** or **–** to move forward or backward in time in steps of 1 day. Turn on the constellation lines, at least, for reference, and watch how Mars appears to move against the background of stars.

HOW DID GALILEO GO BLIND? OR, IT'S THE DATA, STUPID!

Does not use Stellarium.

WHAT'S AN ASTRONOMICAL UNIT?

Does not use Stellarium.

HOW DO WE KNOW THE MASS OF JUPITER, ANYWAY?

To observe the motions of the Galilean satellites:

1. Find Jupiter by clicking on the magnifying glass tool in the menu that pops up vertically from the lower left-hand corner of the screen and type **Jupiter** in the text window that pops up. Hit **Enter**, and the software will center on Jupiter. You will immediately see the Galilean satellites (Ganymede, Io, Callisto, and Europa) labeled near Jupiter. To center on any moon, just click on it.

2. Zoom out and in using the **/** and **** keys until you can see all four Galilean satellites.

3. Use the **]** key or the **[** key to move forward or backward in time in steps of 1 week. Use = or – to move forward or backward in time in steps of 1 day.

To make measurements:

1. Pause the motion at the same time for each day or after a set number of hours, and with a ruler measure the distance on the screen of each satellite from the apparent center of Jupiter.

2. Note the time and date for each measurement you make.

WHAT ARE STARS MADE OF?

Does not use Stellarium.

INFERRING PHYSICAL PROPERTIES

Does not use Stellarium.

READING THE STARS

Does not use Stellarium.

WHAT IS THE MILKY WAY MADE OF?

Does not use Stellarium.

WHERE IN THE MILKY WAY ARE WE?

To view the Milky Way from the Northern and Southern Hemispheres:

1. Set your location in the Northern Hemisphere by clicking on the **View** tab and setting your location in the **Observing Location** window by clicking the **Setup** button.
2. Zoom out (with the – key) and use the mouse to pan around until you see a large band of light—the Milky Way. It will not be labeled.
3. Set your location to a city in the Southern Hemisphere using the same method as in step 1.

To find objects:

1. Click on the **Search** tab.
2. Type in the name of the object in the search window (*Note:* Do not use spaces in the name you type).
3. Click on the icon that appears below the name (even if the name does not match the name you typed in, as long as there is only one icon that appears under the window).
4. Using the mouse, right click on the object to get basic information about the object. </nl>

HOW CLOSE IS OUR NEAREST NEIGHBOR?

Does not use Stellarium.

THE GALAXY ZOO

Does not use Stellarium.

OUR EXPANDING UNIVERSE

Does not use Stellarium.

Instructions for Using the WorldWide Telescope

To become familiar with how to use WWT:

1. Click on the tab **Guided Tours** and view the following tours:
   a. Educators Tours
   b. Interactive Tours
   c. Using the Observing Time Pane Tour

FOR MACS

1. Go to the WWT Web page (www.worldwidetelescope.org).
2. Click on the button **Preview Web Client**.

FOR PCS

1. Go to the WWT Web page (www.worldwidetelescope.org).
2. Click on the button **Install Windows Client** and follow the instructions to install the Windows Client.

The following list relates specific help for Exercises:

MOON PHASES AND SCIENTIFIC MODELS

Does not use WWT.

MOTIONS IN THE NIGHT SKY AND THE CELESTIAL SPHERE

To set the sky to your location:

1. Click on the **View** tab.
2. Click the **Setup** button in the third window from the left. Choose **US Cities** in the pull-down menu under **Data Set**, choose your state and city, and press **OK**.
3. Make sure the window on the bottom under the words **Look At** is set to **Sky**.

To set to the time of your observation:

1. Click on the **View** tab.
2. Adjust the time in the window to match the time of your observation.
3. Use the fast-forward (double arrows) button to see the motion of the stars against the horizon shown.

THE SUN THROUGH THE SEASONS

To set the sky to January 1 of this year:

1. Click on the **View** tab.
2. Click on the pull-down menu where the date is shown.
3. Change the month to 1 by clicking on the up or down arrows next to the window beneath the word **Month**.
4. Change the day to 1 by clicking on the up or down arrows next to the window beneath the word **Day**.
5. Click the **Apply** button.
6. Click the **OK** button.

To observe sunrise:

1. Follow the instructions above to set your location.
2. Click on the **View** tab.
3. Click on the pull-down menus where the date is shown.
4. Change the time to a time near sunrise for January 1 at your location (if you're not sure when sunrise will occur, just pick 6:00:00).
5. Click the **Apply** button.
6. Click the **OK** button.
7. Use the mouse to move around until you are viewing the eastern horizon from your location.
8. Use the fast-forward or rewind buttons to move the sky until you see the Sun (a yellow dot on the blue ecliptic line).

To observe rising constellation at sunset:

1. The name of the constellation in which your cursor is located is shown in the window on the bottom right-hand side of your screen.

MARS'S MOTION AND MODELS OF THE SOLAR SYSTEM

To observe the retrograde motion of Mars:

1. Find Mars by clicking on the image of Mars on the bottom menu bar or by clicking on the tab **Explore**, the picture **Solar System**, then the planet **Mars**.
2. Use the – key to zoom out to a view of the night sky.
3. Click on the **View** tab and use the fast-forward button to move the sky ahead in time. Follow along with Mars as it moves along the ecliptic. The best speed setting for fast-forward is 100000×. Faster than this makes it difficult to move along with Mars, and slower speeds take too long to find the retrograde period.

HOW DID GALILEO GO BLIND? OR, IT'S THE DATA, STUPID!

Does not use WWT.

WHAT'S AN ASTRONOMICAL UNIT?

To view the transit of Venus:

1. Set the location to a city on a single line of longitude at a very high northern latitude.
2. Go to the position of the Sun using the **Explore** tab, click on **Solar System**, then **Sun**.
3. Use the – key to zoom out so you can see the entire disk of the Sun. (If you zoom out too far, use the + key to zoom back in.)
4. Set the date to June 8, 2004, and start the time at 0 hours, 0 minutes, 0 seconds, and use the fast-forward button to see Venus approach the disk of the Sun.

To change location:

1. Set the location to another city on the same line of longitude at a very low southern latitude.
2. Repeat the above steps to make the observations needed for the exercise.

HOW DO WE KNOW THE MASS OF JUPITER, ANYWAY?

To observe the motions of the Galilean satellites:

1. Go to the position of Jupiter using the **Explore** tab, click on **Solar System**, then on **Jupiter**.
2. Zoom out and in using the + and keys until you can see all four Galilean satellites. (You will know that you can see them because they will appear as icons in the lower menu. Hovering your cursor over each image will cause a finder circle and label to appear over each satellite.)
3. To cause the satellites to move relative to Jupiter, use the fast-forward button set to about 1000×.

To make measurements:

1. Pause the motion at the same time for each day or after a set number of hours, and with a ruler measure the distance on the screen of each satellite from the apparent center of Jupiter.
2. Note the time and date (on the **View** tab) for each measurement you make.

WHAT ARE STARS MADE OF?

To find stars:

1. Click on the **Search** tab.
2. Type in the name of the object in the search window. (*Note:* Do not use spaces in the name you type.)
3. Click on the icon that appears below the name (even if the name does not match the name you typed in, as long as there is only one icon that appears under the window).
4. Using the mouse, right-click on the object and click on the **Research** button.
5. Choose **Information** and **Look up on SIMBAD** to get the information you need for this exercise.

INFERRING PHYSICAL PROPERTIES

Does not use WWT.

READING THE STARS

Does not use WWT.

WHAT IS THE MILKY WAY MADE OF?

To find objects:

1. Click on the **Search** tab.
2. Type in the name of the object in the search window. (*Note:* Do not use spaces in the name you type.)
3. Click on the icon that appears below the name (even if the name does not match the name you typed in, as long as there is only one icon that appears under the window).
4. Using the mouse, click on the object to get basic information about the object.
5. Click on the **Research** button and choose **Information** to get the extra information you may need for this exercise.

WHERE IN THE MILKY WAY ARE WE?

To view the Milky Way from the Northern and Southern Hemispheres:

1. Set your location in the Northern Hemisphere by clicking on the **View** tab and setting your location in the **Observing Location** window by clicking the **Setup** button.
2. Zoom out (with the – key) and use the mouse to pan around until you see a large band of light—the Milky Way. It will not be labeled.
3. Set your location to a city in the Southern Hemisphere using the same method as in step 1.

To find objects:

1. Click on the **Search** tab.
2. Type in the name of the object in the search window. (*Note:* Do not use spaces in the name you type.)
3. Click on the icon that appears below the name (even if the name does not match the name you typed in, as long as there is only one icon that appears under the window).
4. Using the mouse, right-click on the object to get basic information about the object.

HOW CLOSE IS OUR NEAREST NEIGHBOR?

Does not use WWT.

THE GALAXY ZOO

To observe galaxies:

1. Click on the **Search** tab.
2. Type in the name of the object in the search window. (*Note:* Do not use spaces in the name you type.)
3. Click on the icon that appears below the name (even if the name does not match the name you typed in, as long as there is only one icon that appears under the window).
4. Using the mouse, right-click on the object to get basic information about the object.

To get more information about galaxies:

1. Click on the **Research** button and choose **Information** to get the extra information you may need for this exercise.

OUR EXPANDING UNIVERSE

Does not use WWT.

Appendix

GEOMETRIC PRINCIPLES FOR CALCULATING DISTANCE OF AN ASTRONOMICAL UNIT

To calculate the distance of an astronomical unit, we need to know SOH CAH TOA:

$\sin \theta$ = opposite/hypotenuse
$\cos \theta$ = adjacent/hypotenuse
$\tan \theta$ = opposite/adjacent

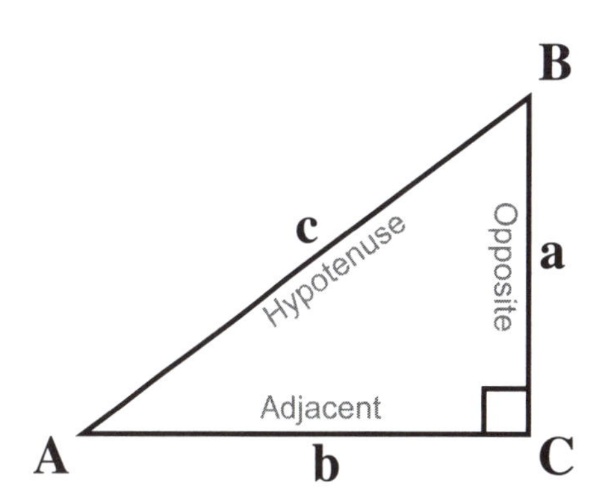

So, for example, to find the length of side b, if you know the angle A and the length of side c, you can find it by setting the cosine of A equal to b/c. Conversely, if you know the sides, but not the angle, you can use the inverse trigonometry functions to figure out the angle.

The inverse trigonometry functions are written as follows:

$\sin^{-1}$(opposite/hypotenuse) = θ
$\cos^{-1}$(adjacent/hypotenuse) = θ
$\tan^{-1}$(opposite/adjacent) = θ

These expressions allow one to calculate an angle, knowing two sides of a triangle.
Another important key to solving this problem is knowing how small angles behave. For small angles, $\tan(\theta/2) = \frac{1}{2}\tan\theta$ and $\sin\theta = \theta$.

Calculating Percent Error

The percent error is the difference between your calculated value and the accepted value divided by the accepted value. So, if *C* is your calculated value and *A* is the accepted value, the equation looks like this:

$(A - C)/A \times 100\%$ = percent error

Note that if *C* is a number larger than *A*, the numerator in the above equation should be *C* – *A*, not *A* – *C*, so that you get a positive number.

Credits

Exercise 1 Opener NASA/JPL

Exercise 2 Opener NASA

Exercise 3 Opener Anthony Ayiomamitis

Exercise 4 Opener Tunc Tezel

Exercise 5 Opener NASA

Exercise 6 Opener NASA

Exercise 7 Opener NASA/Johns Hopkins University Applied Physics Laboratory/Southwest Research Institute/Goddard Space Flight Center

Exercise 8 Opener Michigan State University ISP 205

Exercise 9 Opener NASA, ESA, and The Hubble Heritage Team (STScI/AURA)

Exercise 10 Opener NASA **10.1** Reproduced by permission of the AAS **10.2** AAS **10.3** AAS **10.4** Montgomery, K. A., Marschall, L. A., & Janes, K. A. 1993, AJ, 106, 18, reproduced by permission of the AAS

Exercise 11 Opener NASA/CXC/SAO and Axel Mellinger, University of Potsdam, Germany

Exercise 12 Opener NASA

Exercise 13 Opener NASA, ESA and A. Nota (STScI/ESA)

Exercise 14 Opener NASA, ESA, Hubble Heritage Team (STScI/AURA) Acknowledgment: J. Blakeslee (Washington State University)

Exercise 15 Opener Reproduced by permission of Matthew Colless, Anglo-Australian Observatory